スルメイカ類の
ムチン

－巨大卵塊の形成－

木村　茂

(東京海洋大学名誉教授)

五曜書房

水槽内に産み出されたスルメイカの巨大卵塊
北海道大学臼尻水産実験所の大型水槽内に浮遊する直径約 80 cm に達する球状の透明な卵塊(1995 年 12 月 14 日撮影,桜井泰憲博士のご厚意による)

はじめに

　軟体動物のイカは日本人にとって馴染み深い水産物で，特に沖合性のアカイカ科はスルメイカ類とよばれ，生食用および加工用原料として大量に消費されている．日本列島沿いに生息するスルメイカ（*Todarodes pacificus*）は最も著名な種で，1968年に70万トンもの豊漁に恵まれたが，今は漁獲量の減少が著しい．因みに，2014年における日本のイカ類の漁獲高は20.6万トンで，第1位のスルメイカは16.7万トン（全体の81%）を占め，第2位のアカイカは0.46万トンに過ぎない．かなり以前から，近縁種のアカイカ，アルゼンチンイレックス，ニュージーランドスルメイカなどが市場に出回っており，日本からアメリカ西海岸に至る広い太平洋の沖合に分布する巨体のアカイカは「さきいか」などの加工用として欠かせない．
　ところで，日本海区水産研究所の浜部基次博士は1961～2年にかけて，海底に沈めた木樽の中で飼育しているスルメイカが輸卵管腺に由来する物質に包まれた数千粒の受精卵を含むフットボール大の卵塊を産み，包卵腺の分泌するアルブミン様粘質物が卵塊表層を形成していることを初めて報告した．その後，1986年にNHKテレビの特集「スルメイカの謎を追う」の中で，鳥取県栽培漁業センターの100トン水槽で「花子」と名付けられたスルメイカの産卵が観察され，約20万粒の卵を含む直径1mほどの巨大な卵塊が放映されたという．私は残念ながら見ていないが，イカの研究者にも驚きであったらしい．スルメイカ類は共通して巨大な卵塊を産むと考えられている．
　北海道大学水産学部の桜井泰憲博士らは1994年9月に大型回流水槽で

VI　スルメイカ類のムチン

飼育しているスルメイカの産卵行動の観察に成功し，透明な2個の球形卵塊を入手した．私達は同じ1994年にアルゼンチンイレックスの包卵腺粘質物から単離した糖タンパク質ムチンの構造に関する研究報告をしており，極めて貴重な卵塊を得た桜井博士との間で卵塊の構成物質に関する共同研究を行うことができた．

　本書はムチンとしても特異な構造をもつ包卵腺ムチンが卵塊形成に重要な役割を担っていることをまとめたものである．原生動物，バクテリアなどの外敵から受精卵を守る巨大な卵塊の存在は自然の不思議さを感じさせる．多くの人に馴染みの少ないムチンという魅力的なタンパク質の一端を知っていただければ幸いである．

2016年5月1日　　　　　　　　　　　　　　　　　　　　　　　　　　著者

目　次

はじめに　V

第1章　糖タンパク質ムチンの概要　3
　　1．一般的性質　3
　　2．遺伝子と系統発生　5
　　3．分泌型ムチンの研究例　7

第2章　イカの包卵腺粘質物　9
　　1．化学組成と電気泳動図　10
　　2．新規のアミノ糖：4-O-メチルグルコサミン　15

第3章　包卵腺のムチン　23
　Ⅰ．アルゼンチンイレックス　23
　　1．単離と性質　23
　　2．プロテアーゼ消化　26
　　3．塩溶性のムチン複合体　29
　Ⅱ．スルメイカ　35
　　1．新規の中性糖：4-O-メチルグルコース　35
　　2．単離と性質　36
　　3．主要糖鎖Ⅰの特異な構造　41

第4章　スルメイカの巨大卵塊　53
　　1．卵塊膜のムチン複合体　55
　　2．卵塊内部ゼリーのムチン複合体　60
　　3．卵塊形成と包卵腺粘質物　63

VIII　スルメイカ類のムチン

第5章　包卵腺ムチンの利用　65

　　　　1．ムチンの製造　66
　　　　2．化粧品素材としての性質　66

付録：イカ包卵腺ムチンに関する原著論文　71

索引　84

おわりに　91

スルメイカ類のムチン
－巨大卵塊の形成－

第1章　糖タンパク質ムチンの概要

　動物細胞は脂質二重層およびその中に埋め込まれた糖タンパク質と糖脂質からなる細胞膜によって維持され，外界と接している．呼吸，消化および泌尿生殖の諸器官は内面が非常に軟弱で傷つき易く，粘膜（mucosa）と呼ばれる粘液ゲル層が上皮細胞の表面を覆っており，ある種の両生類（カエルなど）と魚類（ウナギなど）では皮膚の表面にも分布する．この粘膜の主成分がムチン（mucin）で，拡散障壁および潤滑剤として働き，管による物質の輸送促進のみならずバクテリア，ウイルスなどの感染，脱水および物理的あるいは化学的障害から上皮細胞を保護している．それ故，ムチンは体内に侵入した異物を迎え撃つ免疫物質に比べると原始的ではあるが，重要な生体防御物質といえよう．これまでムチンに関する研究は主に哺乳類を対象にした医学的な立場から行われてきた．

1．一般的性質

　ムチンは1本のポリペプチド鎖（コアタンパク質という）のトレオニンまたはセリンの OH 基が，2〜10 個ほどの単糖からなるオリゴ糖鎖の還元末端にある N-アセチルガラクトサミンの C-1 の OH 基と脱水縮合した O-グリコシド結合をもつ粘液性の糖タンパク質である（図1）．細胞の外で機能する分泌型と細胞膜に結合して機能する膜結合型に大別される．ムチンの定義は研究者によって多少の相違があるものの，一般的には「O-グリコシド型（ムチン型）の糖鎖が糖タンパク質分子量の 50％以上を占め，コアタンパク質はプロリン（P），トレオニン（T）およびセリン（S）に富む

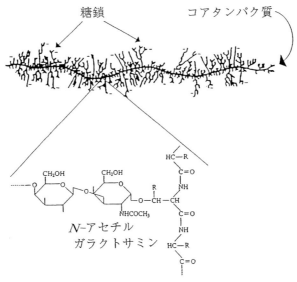

図1 ムチンの模式図とムチン型糖鎖の結合様式
R＝H：セリン，R＝CH₃：トレオニン

PTSドメインが縦に並んだアミノ酸の反復配列構造をもつ」といえる[1]．なお，ドメインとはタンパク質の機能単位で領域と訳される．O-グリコシド型糖鎖は糖タンパク質分野の2大糖鎖群の一つであるアスパラギンに結合したN-グリコシド型糖鎖が糖タンパク質分子量の10%以下に過ぎないのとは対照的である．

多様な構造をもつムチンの糖鎖は一般に還元末端にあるN-アセチルガラクトサミンの他にN-アセチルグルコサミン，ガラクトース，フコースおよびシアル酸の5種類の単糖から構成され，シアル酸を含むシアロムチンとシアル酸の代わりに硫酸基を含むスルホムチンに分類される．ムチンのO-グリコシド結合は希アルカリ(希薄なアルカリ溶液)に対して不安定で，糖鎖が0.05 M KOH-45℃-15 hの処理でβ離脱反応を起こしてコアタンパク質から離れ，ムチンは本来の性質を失う．逆に，糖鎖の切り出しに

は希アルカリ処理が有効である．ムチンは動物のみならず植物，例えばヤマイモの粘質物に O-グリコシド結合をもつムチン様の糖タンパク質が存在する[2]．なお，ムチンという単語は海藻などのネバネバしたムコ多糖類などに対しても一般的な意味で用いられるので注意したい．

　分泌型ムチンは一般にゲル形成能をもち，オリゴ糖鎖が密に結合したPTSドメインは二次構造を欠いて長く伸びた形となり屈曲性に乏しい．すなわち，オリゴ糖鎖はペプチド結合の周囲の回転を制限し，隣接する負荷電（シアル酸または硫酸基）によって生じる反発力のためにムチン分子の半剛性に寄与している．さらに，オリゴ糖鎖はタンパク質分解酵素がコアタンパク質のペプチド結合に接近することを妨げ，酵素分解に対する抵抗性を増加させる．このような長く伸びたゲル形成ムチンは高粘性の水溶液を生じ，NあるいはC末端のシステイン含有ドメインによるS-S結合により，多量体の形成が可能となる．なお，分泌型の非ゲル形成ムチンは多量体を形成せず，その構造と機能は不明な点が多い．一方，膜結合型ムチンは多量体を形成しないが，膜貫通ドメイン，細胞接着ドメインなど数種の特徴的なドメインをもち，多くは上皮細胞（頂端側）の膜に存在して情報伝達を担っている．ある種の膜結合型ムチンはガン（癌）の進行に関与するらしい．

2．遺伝子と系統発生

　ムチンのコアタンパク質（MUC：マック）をコードする遺伝子群は*MUC*遺伝子ファミリーと呼ばれる．脊椎動物ムチンの場合，発見順にアラビア数字を付けて区別し，ファミリーに属する多くのタンパク質（MUC1～22）が報告された[3]．ヒトでは分泌型と膜結合型を合わせて次の21種である．なお，MUC5BとMUC9並びにMUC11とMUC12は各々，同一のタンパク質と判明したので，MUC9およびMUC12を削除した．

分泌型：1) ゲル形成……MUC2/5AC/5B/6/19 の 5 種
　　　　2) 非ゲル形成……MUC7/8 の 2 種
膜結合型：MUC1/3A/3B/4/11/13/14/15/16/17/18/20/21/22 の
　　　　14 種

　ところで，ウシでは分泌型 6 種と膜結合型 9 種の合計 15 種が同定され，ヒトに存在する MUC3B/8/17 は同定されていない．研究の進んでいるウシ顎下腺ムチン(BSM)は分泌型の MUC19 の一部とする報告がある．

　すべてのムチンは多数のオリゴ糖鎖が結合した PTS ドメインで特徴付けられる．PTS ドメインは一般にアミノ酸 20 残基ほどの比較的短い反復配列が認められ，最短と思われるヒト唾液ムチン(MUC7)のドメインはアミノ酸 23 残基が 6 回反復した 138 残基からなる[4]．一方，魚（ゼブラフィッシュ）の MUC5 はアミノ酸 2,571 残基からなる配列が 61 回も反復され，15 万残基以上の著しく長い PTS ドメインが知られている[5]．なお，タンパク質の構成アミノ酸および糖鎖の構成単糖は各々，脱水縮合して結び付いており，1 分子の水がとれた残りという意味からアミノ酸残基および糖残基という．

　1990 年頃から生物情報科学の急速な発達によって，ゲノム配列情報からムチン遺伝子の検出・同定が可能となったが，PTS ドメインはアミノ酸配列の多様性のゆえに進化学的分析には不適当である．Lang らはムチンにも特徴的なフォンビルブランド D(VWD)ドメインおよびウニ精子 63 kDa タンパク質，エンテロキナーゼ，アグリンなどに共通する SEA ドメインを併用して，下等なイソギンチャクからヒトに至る 12 種の多細胞動物のゲノム配列を解析し，ムチンの系統発生に関する研究を行った[5]．その結果，ゲル形成ムチンは多細胞動物進化の初期（刺胞動物）にはすでに出現しており，その起源はかなり古いことが判明した．カエルはゲル形成ムチンの種類が著しく多く，哺乳類の 5 種に対して，少なくとも 25 種に達する点が際立っている．魚類はゼブラフィッシュに 1 種の MUC2 と 6 種

のMUC5が同定された．一方，膜結合型のムチンはいずれも脊椎動物になって初めて出現し，脊椎動物に特有の情報伝達および制御機構に関係している．

3．分泌型ムチンの研究例

　糖タンパク質の分泌型ムチンを単離した例は比較的少ないが，哺乳類の胃粘液[6]，顎下腺，気管支腺などのムチンは医学的な生体防御に関連して良く研究され，報告例は多い．魚類ではウナギ[7]，ドジョウ[8]，ニジマス[9]などの皮膚粘液ムチン（すべてシアロムチン）が報告された．無脊椎動物では本書に述べるスルメイカ類の包卵腺ムチンの他に，タイラギ（二枚貝）の真珠層ムチン[10]，クラゲ類のムチン[11]などで，研究例は著しく少ない．

　分泌型ムチンは一般に分子量が100万以上の巨大な多量体を形成し，糖鎖の構造が不均一なうえにシアル酸あるいは硫酸基という酸性基が多く，その単離は現在の進歩した分離技術をもってしても容易でない．通常は粘膜または粘液をホモジナイザーで懸濁液としてプロテアーゼ阻害剤，グアニジン塩酸，尿素，還元剤，界面活性剤などを適宜に含む緩衝液でムチンを可溶化し，遠心分離または濾過で残渣を除き，ゲル濾過，エタノール沈殿などで高分子量画分のムチンを分離する．さらに，必要に応じて密度勾配遠心法でムチンを精製する．

【文　献】

1 ）Strous, G. J. and Dekker, J.: *Cri. Rev. Biochem. Mol. Biol.*, **27**, 17520（1992）
2 ）津久井 学：関東学院大学人間環境研究所所報，**5**, 7（2007）
3 ）Hoorens, P. R., Rinaldi, M., Li, R. W., Goddeeris, B., Claerebout, E., Vercruysse, J. and Geldhof, P.: *BMC Genomics*, **12**, 140（2011）
4 ）Bobek, L. A., Tsai, H., Biesbrock, A. R. and Levine, M. J.: *J. Biol. Chem.*, **268**, 20563（1993）

5) Lang, T., Hansson, G. C. and Samuelsson, T.: *Proc. Natl. Acad. Sci. U.S.A.*, **104**, 16209 (2007)
6) 堀田恭子, 石原和彦：胃粘液の魅力を探る 最新手法によるムチンの解明, メジカルビュー社, pp. 1 ～ 97 (1999)
7) Sumi, T., Hama, Y., Nakagawa, H., Maruyama, D. and Asakawa, M.: *J. Fish Biol.*, **64**, 100 (2004)
8) Kimura, M., Hama, Y., Sumi, T., Asakawa, M., Rao, B. N. N., Horne, A. P., Li, S-C., Li, Y-T. and Nakagawa, H.: *J. Biol. Chem.*, **269**, 32138 (1994)
9) Sumi, T., Hama, Y., Maruyama, D., Asakawa, M. and Nakagawa, H.: *Biosci. Biotech. Biochem.*, **61**, 675 (1997)
10) Marin, F., Corstjens, P., de Gaulejac, B., de Vrind-De jong, E. and Westbroek, P.: *J. Biol. Chem.*, **275**, 20667 (2000)
11) Masuda, A., Baba, T., Dohmae, N., Yamamura, M., Wada, H. and Ushida, K.: *J. Nat. Prod.*, **70**, 1089 (2007)

第2章 イカの包卵腺粘質物

　包卵腺はこれまでにもイカの性成熟度を示す一つの指標として，長さと重さが測定されているものの，内部の粘質物は全く顧みられなかった．図2に示すスルメイカ（雌）の解剖模式図を見ると，まるで魚の白子（精巣）に似た"笹かまぼこ"形の一対の器官がまず目に留まる．これが産卵時に大きな卵塊を形成するための粘液を分泌する包卵腺または纏卵腺（てんらんせん）と呼ばれる器官である．私が初めて包卵腺と出会ったのは，（株）紀文食品の加藤 登博士から「アルゼンチン沖で大量に水揚げされるアルゼンチンイレックス（アカイカ科の一種：通称は松いか）の包卵腺粘質物をねり製品に利用したいので，成分を調べてほしい」との依頼を受けた1990年頃であった．包卵腺は長さ10 cmで重さ30 gほどのたらこ状の形をして，つきたての餅のような粘質物で満たされている．これまで，包卵腺は不可食部とみなされることが多かった．

　ところで，山東料理に鍋粑烏魚蛋（グオバァウユィダン）という"イカの内卵のお焦げ"と称される料理がある[1]．中国語の烏魚蛋とはイカの包卵腺のことで，含有する粘質物を茹でてショウガや老酒を用いて臭いを抜き，調味用と合わせて食べる料理である．味は特に無いので，茹でた卵白様の食感を楽しむ料理と思われる．中国の山東地方では塩漬けした保存食品として利用しているらしい．最近，生食（刺身）用として，魚肉冷凍すり身に包卵腺粘質物を加えて従来のねり製品にない"ふんわり"とした食感をもつ「イカ作り」と称する製品が市販されたことがあり，好評を得ていた．私も食べてみたが，ねり製品として独特の優れた味わいがある．このよう

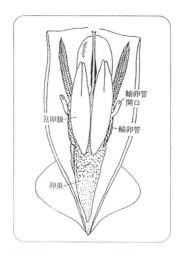
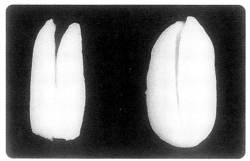

図2　スルメイカ(雌)の解剖模式図と包卵腺

に，包卵腺粘質物は一部で食用にされているが，化学的研究は皆無である．
　以下では粘質物の化学組成とタンパク質組成を調べた．

1．化学組成と電気泳動図[2,3]

　試料に用いた包卵腺は表1に示すアカイカ科3種およびジンドウイカ科1種から分離し，コラーゲン性の薄皮に包まれた粘質物を分析に供した．

表1 イカ類の包卵腺

	重さ(g)	長さ(cm)	幅(cm)
アカイカ科			
スルメイカ　*Todarodes pacificus*	11	7	2.5
アルゼンチンイレックス　*Illex argentinus*	30	12	4.5
アカイカ　*Ommastrephes bartramii*	80	15	4.5
ジンドウイカ科			
アオリイカ　*Sepioteuthis lessoniana*	22	7	3.5

初めにアルゼンチンイレックスを試料に一般組成を分析した[2]。粘質物はかなり多量のアミノ糖を含むので，通常の分析法を次のように変更した．

(1) 粗タンパク質はケルダール法で測定した全窒素からアミノ糖由来の窒素を差し引き，6.25を掛けて算出した．

(2) 糖質はオルシノール-硫酸法で定量した中性糖に，高速液体クロマトグラフ(HPLC)で定量したアミノ糖を加えて算出した．アミノ糖は粘質物を加水分解(4 M HCl-100℃-8 h)の後に，市販のアミノ酸分析用 0.35 M クエン酸緩衝液(pH 6.1)で溶出し，ニンヒドリン法で定量した．

粘質物の一般組成は乾重量当り粗タンパク質 48.6％，糖質 37.1％，脂質 6.9％，灰分 5.5％および特殊成分の硫酸基 2％である．なお，アミノ糖の分析では思いもかけず，グルコサミン(GlcN)とガラクトサミン(GalN)の間に未知のアミノ糖 X が検出された(図3)．この X 成分は GlcN と同じニンヒドリン発色率をもつと仮定して定量したが，次節に述べる新規の 4-*O*-メチルグルコサミンと同定された．

次にアミノ酸，中性糖およびアミノ糖の組成を分析し，1000 残基中の残基数として表2に示す[3]．中性糖は粘質物を 2.5 M トリフルオロ酢酸(TFA)で加水分解(100℃-7 h)の後，TFA 化した糖アルコールをガスクロマトグラフで分析した．アミノ酸・糖組成はイカの種類で多少異なるが，

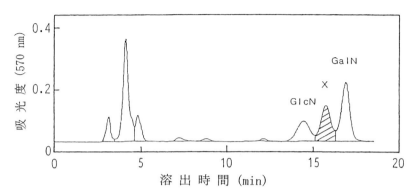

図3　アルゼンチンイレックス包卵腺粘質物の未同定アミノ糖X
試料を加水分解(4 M HCl-100℃-8 h)の後に，HPLCのアミノ酸分析システム(東ソー)で分析した.
カラム：TSKgel Aminopak(55℃)
溶出液：0.35 M クエン酸緩衝液(pH 6.1)
流速：0.4 ml/min，検出：ニンヒドリン法
GlcN：グルコサミン，GalN：ガラクトサミン

残基数としてアミノ酸67〜80%，中性糖13〜23%およびアミノ糖7〜15%である．主要成分はアカイカ科の3種に共通するガラクトースとトレオニンで，これらの他にスルメイカのフコースが認められた．一方，ジンドウイカ科のアオリイカは主要成分がアスパラギン酸とグルタミン酸で，アカイカ科とかなり相違している．各種のアミノ酸はほぼ平均的に分布しているが，エキス由来のタウリンは2〜4%が検出された．注目の4-O-メチルグルコサミンは全てのイカに検出されたが，アオリイカはアカイカ科に比べて少なかった．

　比較のために分析したスルメイカ輸卵管腺(重さ：0.45 g，直径：1 cm)の粘質物はアミノ酸86%，中性糖7.4%およびアミノ糖6.6%を含み，糖質の割合が著しく少ない．主要アミノ酸はグリシン，シスチン/2およびリシンの3種で，包卵腺粘質物のそれとは大きく異なる．

表2 イカ類の包卵腺粘質物と輸卵管腺粘質物のアミノ酸・糖組成(残基数/1000)

	包卵腺				輸卵管腺
	アカイカ科			ジンドウイカ科	アカイカ科
	スルメイカ	アルゼンチンイレックス	アカイカ	アオリイカ	スルメイカ
アミノ酸					
タウリン	18	26	30	43	18
アスパラギン酸	63	65	68	75	65
トレオニン	82	74	82	54	69
セリン	31	28	28	49	36
グルタミン酸	51	54	63	77	65
プロリン	55	57	56	44	48
グリシン	60	51	53	52	113
アラニン	30	34	38	40	38
シスチン/2	40	23	25	15	80
バリン	35	36	36	46	35
メチオニン	7	8	9	14	6
イソロイシン	51	48	58	51	38
ロイシン	25	35	32	64	17
チロシン	25	21	19	29	24
フェニルアラニン	23	21	23	39	43
トリプトファン	−	3	−	−	−
リシン	56	51	53	52	115
ヒスチジン	14	14	12	16	8
アルギニン	27	22	28	35	42
小計	693	671	713	795	860
中性糖					
ラムノース	5	24	5	1	0
フコース	70	59	23	54	21
リボース	3	13	5	7	3
キシロース	7	24	11	4	15
マンノース	7	19	6	14	3
グルコース	13	20	5	3	2
ガラクトース	83	72	82	51	30
4-O-メチルグルコース[*1]	−	−	−	−	−
小計	188	231	137	134	74
アミノ糖[*2]					
グルコサミン	42	19	26	29	32
4-O-メチルグルコサミン	19	36	68	5	5
ガラクトサミン	58	43	56	37	29
小計	119	98	150	71	66
合 計	1000	1000	1000	1000	1000

*1：分析を行った1994年当時，4-O-メチルグルコースの存在は不明．
*2：天然にはアミノ基がアセチル化している．

14　スルメイカ類のムチン

　最後に，包卵腺粘質物のタンパク質組成を 0.1%SDS(ドデシル硫酸ナトリウム)を含む 3.5%ポリアクリルアミドゲル電気泳動法(以後，SDS-ゲル電気泳動と略す)で調べた[3]．試料を 1%SDS-4 M 尿素-0.02 M リン酸緩衝液(pH 7.2)に溶かして還元剤 DTT(ジチオトレイトール)で処理する

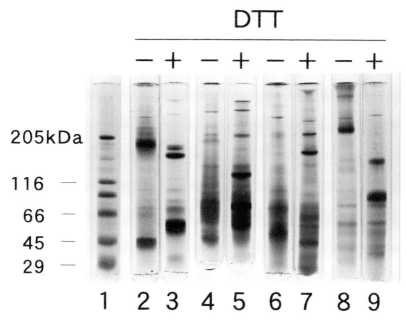

図4　イカ類包卵腺粘質物の SDS-ゲル電気泳動図
　3.5％ポリアクリルアミドゲルを用い，試料を 0.1%SDS-3.5 M 尿素-0.02 M リン酸緩衝液(pH7.2)で泳動し，クマーシーブリリアントブル-R-250 で染色した．
　1：標準タンパク質
　2, 3：スルメイカ
　4, 5：アカイカ
　6, 7：アルゼンチンイレックス
　8, 9：アオリイカ
　DTT：ジチオトレイトール(還元剤)

前後の泳動図を図4に示す．粘質物のタンパク質成分はイカの種類によって明らかに異なり，ゲル上面の高分子量成分の他に，多数の成分が分子量3～22万の間に検出された．泳動図はDTTでS-S結合を還元すると大きく変化し，ゲル上面の高分子量成分が明らかに減少していた．

2．新規のアミノ糖：4-*O*-メチルグルコサミン[4]

アルゼンチンイレックス包卵腺粘質物の加水分解物に存在する未知のアミノ糖Xを単離して構造を解析し，これまで自然界に存在が知られていない新規の4-*O*-メチルグルコサミンであることを明らかにした．

粘質物4g(湿重量)を400mlの4M HClで加水分解(100℃-8h)を行い，少量の活性炭で処理した後に乾固して蒸留水に溶解した．構造解析に必要な量を調製するため，加水分解物40mgを50℃に保ったスルホン化ポリスチレン樹脂のアミノ酸分析用カラムに添加し，0.33M HClで溶出した[5]．溶出液はElson-Morgan反応を改良したBlix法[6]でアミノ糖を検出し(図5)，溶出順にグルコサミンとガラクトサミンに続く第3のアミノ糖X 1.2mgを単離した．

アミノ糖Xの構造は高速原子衝撃-質量分析法(FAB-MS)と核磁気共鳴法(NMR)を用いて解析した．陽イオンモードのFAB-MSの測定はJEOL JMS SX-102マススペクトロメータ並びに^{1}Hおよび^{13}C NMRの測定はJEOL JNM α-500またはBruker AC-300PのNMRスペクトロメータを使用した．^{1}Hおよび^{13}C NMRの化学シフトは各々，重水(δ_H 4.63)およびジオキサン(δ_C 67.8)を標準物質に用いた．

まず，アミノ糖Xはマトリックスとしてグリセロールを用いるFAB-MSによって，[M+H]$^+$に対してm/z 194にピークが認められ(図6-A)，その分子量は193である．さらに，グリセロール-d_3を用いるFAB-MSでは[(M−d_5)+D]$^+$に対してm/z 200にピークがあり，5個の交換可能なプロトンの存在が判明した(図6-B)．

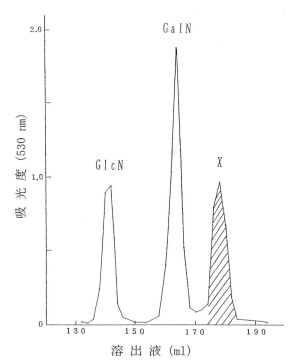

図5　アミノ糖Xの単離
　　　粘質物の加水分解物 40 mg をアミノ酸分析用の液体クロマトグラフ (JLC-3BC)で分析した.
　　　カラム：0.9×70 cm (50℃)
　　　溶出液：0.33 M HCl
　　　流速：0.53 ml/min
　　　検出：Blix 法

　次に，重水中で測定したアミノ糖 X の ^{1}H および ^{13}C NMR データ(表3)はアミノ糖の特徴を良く示している. ^{1}H NMR スペクトルから，α と β のアノマーが 2:1 で存在していた. アノメリックプロトンシグナルの結合定数 J_{HH} 値から，主要なアノマーは α (δ5.32, J = 3.5 Hz)で，β (δ4.80,

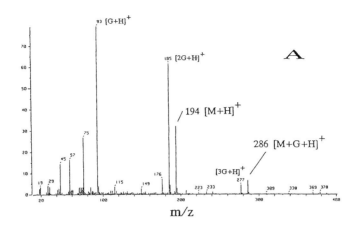

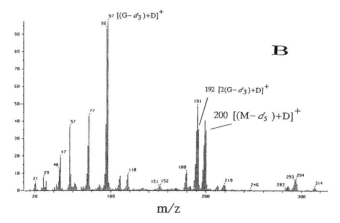

図6 アミノ糖XのFAB-MSスペクトル(陽イオンモード)
マトリックス：(A)グリセロール，(B)グリセロール-d_3

$J=8.5$ Hz)はマイナーである．^{13}C NMRでは2セットのアノマーからなる14のシグナルがあり，各アノマーは7つのシグナルをもっていた．さらに，DEPT ^{13}C NMRスペクトルは14のシグナルを示し，メチル基2個，メチレン基2個およびメチン基10個の炭素が検出された．^{13}Cの化学シフトの値からは各アノマーの1つのC原子(δ55.5, α-C-2：δ58.0, β-C-2)

表3 アミノ糖Xの重水中における ^{1}H および ^{13}C NMR データ

位置	^{1}H	^{13}C	HMBC 相関
α-アノマー			
1	5.32(d 3.5)	90.2(d)	
2	3.20(dd 10.7, 3.5)	55.5(d)	3
3	3.87(dd 10.7, 9.2)	70.7(d)	1, 2
4	3.18(dd 10.0, 9.2)	80.5(d)	5, 6, O-Me
5	3.75(m)	71.9(d)	1, 4, 6
6	3.66(m)	61.3(t)	
	3.72(m)		
O-Me	3.48(s)	61.6(q)	
β-アノマー			
1	4.80(d 8.5)	93.9(d)	
2	2.91(dd 10.7, 8.5)	58.0(d)	1, 3
3	3.67(dd 10.7, 8.8)	73.0(d)	2, 4
4	3.18(dd 10.0, 8.8)	80.6(d)	6, O-Me
5	3.39(ddd 10.0, 5.4, 2.3)	76.5(d)	4, 6
6	3.64(dd 12.3, 5.4)	61.4(t)	
	3.78(dd 12.3, 2.3)		
O-Me	3.48(s)	61.7(q)	

がN原子に結合し，他の6つのC原子はO原子が結合している．これらの結果から，アミノ糖Xの分子式は $C_7H_{15}NO_5$ と判明した．

α-アノマーの ^{1}H-^{1}H COSY, HMQC[7] および HMBC[8] の各スペクトル（本章末に資料として一部を示す）による拡張二次元NMR分析からは図7-Aの平面構造が決定された．解析の詳細は省略するが，これらのデータから，アミノ糖Xは 4-O-メチル-α-グルコサミンと同定され（図7-B），その絶対的立体配置はD型であることが $[\alpha]_D = +82.1°$ (c 0.3, H_2O) から判明した．この値は合成品に対する文献値の $[\alpha]_D = +90.2 \pm 2°$ (c 1.7, H_2O) と良く似ている[9]．なお，β-アノマーはC-1の相対的立体配置を除いて，α-アノマーと同一構造であった．

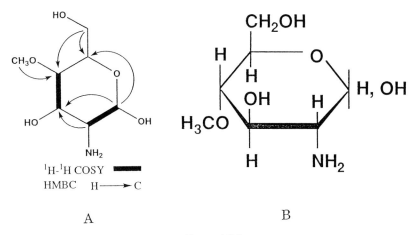

図7 アミノ糖 X の構造
A：α-アノマーの ^{1}H-^{1}H COSY および HMBC 相関
B：4-O-メチルグルコサミン

　この 4-O-メチル-D-グルコサミンはアルゼンチンイレックスの他にスルメイカ，アカイカおよびアオリイカの包卵腺粘質物とスルメイカの輸卵管腺粘質物にも見出された．さらに，粘質物の加水分解物（2 M HCl-100℃-2 h）は等モルの酢酸とアミノ糖を含み，このアミノ糖は天然状態でアミノ基がアセチル化した N-アセチル-4-O-メチルグルコサミンとして存在することが判明した．N-アセチル化したアミノ糖の系統名は次のごとく 2-(acetylamino)-2-deoxy-4-O-methyl-D-glucose である．脱アセチル化したアミノ糖の系統名を 2-amino-2-deoxy-4-O-methyl-D-glucose として東京大学の CASTOR で検索したが，同一化合物は存在しなかった．このメチル化糖は多分，初めて天然に見出されたユニークな単糖である．ただし，methyl の位置が決まっていない化合物は多数あり，2-amino-2-deoxy-3-O-methyl-D-glucose という化合物も認められた．後述するように，包卵腺ムチンは 4-O-メチルグルコースを含み，さらに，線虫の O-

グリコシド型糖鎖には 2-O-メチルフコースと 4-O-メチルガラクトースの存在が知られている[10]. メチル化糖は無脊椎動物に広く分布しているらしく, 今後の比較生化学的研究が待たれる.

【文 献】

1) 専門料理：26 巻, 8 号, 203(1991)
2) Kimura, S., Sugiura, Y., Mizuno, H., Kato, N. and Hanaoka, Y.: *Fisheries Sci.*, **60**, 193 (1994)
3) 木村 茂：平成 8～9 年度科学研究費補助金(基盤研究(B)(2))：研究課題番号 08456103)の研究成果報告書(1998)
4) Kimura, S., Tsunoo, T., Matsuda, H., Murakami, M. and Yamaguchi, K.: *Fisheries Sci.*, **65**, 325 (1999)
5) Crumpton, M. T.: *Biochem. J.*, **72**, 479 (1959)
6) Blix, G.: *Acta Chem. Scand.*, **2**, 467 (1948)
7) Bax, A. and Subramanina, S.: *J. Magn. Reson.*, **67**, 565 (1986)
8) Bax, A. and Summers, M. F.: *J. Am. Chem. Soc.*, **108**, 2093 (1986)
9) Jeanloz, R. W. and Gansser, C.: *J. Am. Chem. Soc.*, **79**, 2583 (1957)
10) Khoo, K. -H., Maizels, R. M., Page, A. P., Taylor, G. W., Rendell, N. B. and Dell, A.: *Glycobiology*, **12**, 163 (1991)

第2章 イカの包卵腺粘質物

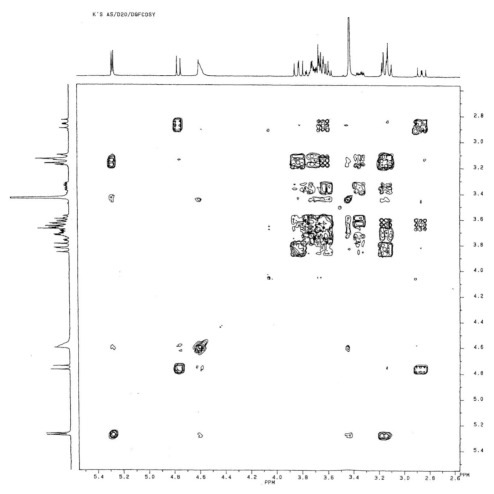

資料1 アミノ糖X(α-アノマー)の $^1H-^1H$ COSYスペクトル

22 　スルメイカ類のムチン

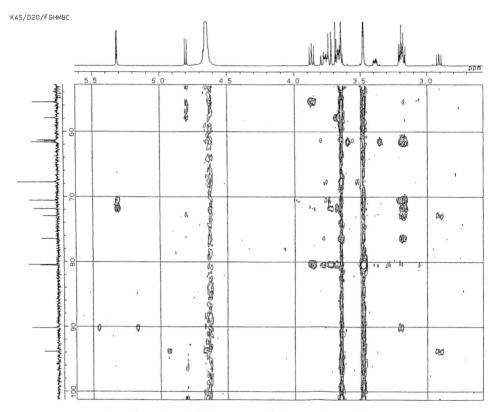

資料2　アミノ糖X(α-アノマー)のHMBCスペクトル

第3章 包卵腺のムチン

　包卵腺粘質物は産卵行動の観察から，卵塊表層のゲル状膜(卵塊膜)を形成すると言われている．本章ではアルゼンチンイレックスとスルメイカの包卵腺粘質物を試料に，粘性を担うムチンを可溶化の後に単離して性質を調べた．

I．アルゼンチンイレックス[1,4]

1．単離と性質[1]

　塩釜の魚市場で購入した包卵腺は平均 35 g で，解凍後にコラーゲン性薄皮をハサミで切って粘質物を取り出した．この粘質物 3 g (水分 76%)を蒸留水 150 ml と共にホモジナイザーで均質な懸濁液とし，これに 7.5 M NaOH を撹拌しながら少量ずつ添加した．粘質物は NaOH 濃度 0.1 M あたりから急激に粘性と透明性を増して溶け始めた．そこで，希アルカリ処理(0.4 M NaOH-4℃-4 h)による可溶化で生じた粘稠な溶液を遠心分離の後に，最終濃度 50% になるようにエタノール(0.2 M NaCl を含む)を加えると，白色の繊維状沈殿(可溶化ムチン)が得られた．このエタノール沈殿を 60% エタノールで洗浄して蒸留水に溶かし，大量の水に対して透析の後に凍結乾燥した．その収量は粘質物(無水物)の約 35% であった．

　アミノ酸と単糖の合計を 1000 残基とした時の残基数で示した化学組成はアミノ酸 34.1% および単糖 65.9% で，硫酸基が単糖の 10.9% に結合している(表4)．アミノ酸はトレオニン，プロリンおよびイソロイシンに著

24　スルメイカ類のムチン

表4　アルゼンチンイレックス包卵腺ムチンの化学組成(残基数/1000)

	包卵腺ムチン	
	全体	プロテアーゼ抵抗性断片
アミノ酸		
アスパラギン酸	13	3
トレオニン	125	132
セリン	4	1
グルタミン酸	10	3
プロリン	65	68
グリシン	10	5
アラニン	6	1
シスチン/2	0	0
バリン	11	6
メチオニン	1	1
イソロイシン	63	65
ロイシン	8	2
チロシン	3	1
フェニルアラニン	4	1
リシン	11	4
ヒスチジン	2	1
アルギニン	5	2
小計	341	296
中性糖		
ラムノース	52	62
フコース	122	107
リボース	16	26
キシロース	35	49
マンノース	6	7
グルコース	34	14
ガラクトース	167	190
4-O-メチルグルコース[*1]	—	—
小計	432	455
アミノ糖[*2]		
グルコサミン	33	31
4-O-メチルグルコサミン	78	88
ガラクトサミン	116	130
小計	227	249
合　計	1000	1000
硫酸基[*3]	72	81

*1：分析を行った1994年当時，4-O-メチルグルコースの存在は不明．
*2：天然にはアミノ基がアセチル化している．
*3：硫酸基の結合している糖の残基数．

しく富み，そのモル比は約 2：1：1 である．これら 3 種のアミノ酸は全アミノ酸の 74.2% を占めていた．単糖はガラクトース，フコース，N-アセチルガラクトサミンおよび N-アセチル-4-O-メチルグルコサミンが主成分で，その他に N-アセチルグルコサミン，ラムノース，キシロースなどが認められた．なお，S-S 結合を形成するアミノ酸のシスチン/2 およびシアロムチン由来のシアル酸は検出されなかった．この化学組成は包卵腺から単離した繊維状物質が糖タンパク質のスルホムチンであることを示唆しており，N-アセチルガラクトサミン（GalNAc）の 116 残基とトレオニン（Thr）の 125 残基はほぼ同数で，脊椎動物のすべてのムチンに存在する GalNAc-Thr 間の O-グリコシド結合が予想される．この O-グリコシド結合は希アルカリ処理で β 離脱反応を起こして開裂し，Thr が破壊される．しかし，包卵腺由来のムチンは希アルカリ処理で Thr 含量に変化がなく，しかも GalNAc を糖鎖の還元末端にもつムチンに発色する Morgan-Elson 反応の変法[2]で発色せず，β 離脱反応が起きていないことは明らかである．それゆえ，GalNAc-Thr 間の O-グリコシド結合は考えられない．

なお，単糖とエステル結合している硫酸基は 1240 cm^{-1} と 820 cm^{-1} に赤外吸収が認められ，エクアトリアル結合をしている（図 8）．

可溶化した包卵腺ムチンは水に溶けて粘稠な溶液を生じ，SDS-ゲル電気泳動でゲル上面に留まる高分子量成分である（図 9）．希薄な 0.025% 水溶液の還元粘度を測定した結果を図 10 に示す．NaCl 非存在下のムチン水溶液は 100 dl/g という著しく高い粘性をもつが，NaCl の添加で急激な粘度低下をきたし，0.02 M NaCl では 17 dl/g になった．NaCl 非存在下の高い粘性は硫酸イオン（SO$_4^{2-}$）の負荷電間の強い反発力によって，伸張したムチン分子の相互作用が著しく増すためといえよう．なお，ムチン水溶液は NaCl の存在でわずかに濁りを生じ，ムチン分子の一部が凝集しており，正確な粘度測定が出来なかった．

以上の結果，包卵腺ムチンは脊椎動物ムチンとは対照的に GalNAc-Thr

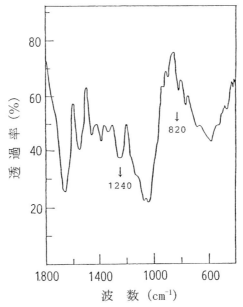

図8 アルゼンチンイレックス包卵腺ムチンの
赤外吸収スペクトル
矢印：硫酸基と単糖のエステル結合を示
す波数.

間の O-グリコシド結合を欠き，分泌型のゲル形成ムチンに認められる S-S 結合による多量体を形成しないなど，極めて特異な性質をもつことが判明した．

2．プロテアーゼ消化[1]

 ムチンは一般にプロテアーゼ(タンパク質分解酵素)に対して抵抗性を示す．そこで，包卵腺ムチンを放線菌 *Streptomyces griseus* の産生する中性プロテアーゼのアクチナーゼ E(科研製薬)で処理してみた．ムチン 200 mg を 10 mM $CaCl_2$ と 3%エタノールを含む 200 ml の 0.5 M 酢酸ナトリウ

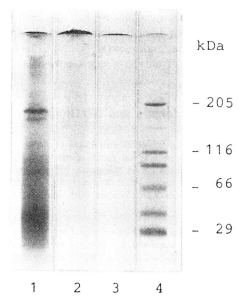

図9 アルゼンチンイレックス包卵腺の粘質物とムチンのSDS-ゲル電気泳動図
分析条件は図4と同じ.
1:粘質物
2:粘質物由来のムチン
3:ムチンのプロテアーゼ抵抗性断片
4:標準タンパク質

ム(pH 8.0)に溶かし,プロテアーゼ 4 mg を加えて 50℃-22 h の消化を行った.反応終了の後,最終濃度 0.1 M になるように NaOH を 4℃で加えてプロテアーゼを失活させ,プロテアーゼ抵抗性断片を 50%エタノールで沈殿させた.この結果,ムチンの約 85%に相当する高分子量断片が沈殿となり,低分子量断片は 50%エタノールに可溶であった.以下では高分子量断片をプロテアーゼ抵抗性断片と呼ぶ.

この抵抗性断片は 0.6 M NaCl を含む 0.02 M 酢酸ナトリウム(pH 7.2)に

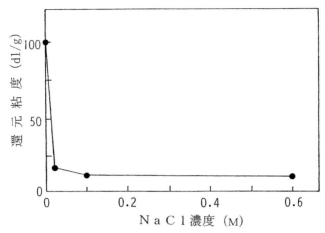

図10　アルゼンチンイレックス包卵腺ムチンの還元粘度に及ぼす NaCl の影響
測定温度：25℃
ムチン濃度：0.025%
溶媒：水

完全に溶けて粘稠な溶液を生じるので,粘度と沈降速度の測定を行った.還元粘度を濃度0%に外挿した固有粘度 $[\eta]$ は 14.7 dl/g で（図11),濃度0.045%における沈降係数 s は 7.1 S(S：スベドベリ単位)である.この断片の分子形を剛体の偏長回転楕円体と仮定して $[\eta]$ の値を Simha の式[3]に代入すると,軸比(長軸と短軸の比)は210と算出された.このように著しく細長い形状の断片は単一の鋭いピークをもつ沈降速度図を示し(図12),SDS-ゲル電気泳動で3.5%ゲルに入れないほどの高分子量成分である(図9).

次に,化学組成はコアタンパク質29.6%および糖質70.4%で,硫酸基が単糖の11.5%に結合している(表4).消化前に比べて,コアタンパク質の割合は4.5%少ない.トレオニン,プロリンおよびイソロイシンのモル比は約2:1:1で,これら3種アミノ酸の合計は全アミノ酸の実に90%

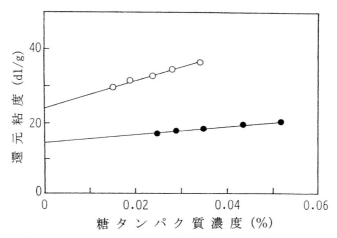

図11 アルゼンチンイレックス包卵腺ムチンのプロテアーゼ
抵抗性断片の固有粘度
○：溶媒　水，$[\eta] = 24.0$ dl/g
●：溶媒　0.6 M NaCl＋0.02 M リン酸緩衝液(pH 7.2)，
　　$[\eta] = 14.7$ dl/g

を占め，トレオニンのみでも45％に達する．このような電気泳動図とアミノ酸組成の特徴から，ムチンの他に糖タンパク質の混在は考え難い．

以上の結果，プロテアーゼは糖鎖の比較的少ない領域，すなわち，包卵腺ムチンの15％ほどを占める分子末端部を消化し，糖鎖の密に結合した分子中央部はプロテアーゼに抵抗性をもつことが明らかになった．私達は末端部を除去したムチンをアテロムチン(atelomucin)と呼んでいる．因みに，aは無を，teloは末端を意味する接頭語である．このアテロムチンは水に溶かすと，高い透明性を示した．

3．塩溶性のムチン複合体[4]

包卵腺粘質物はスルホムチンの他に多種類のタンパク質を含むので，水溶性，塩溶性および不溶性に3分画した．粘質物3gを蒸留水150 ml と共

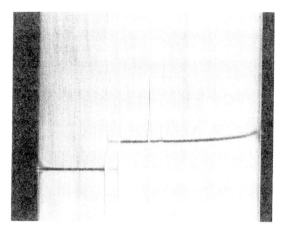

図12 アルゼンチンイレックス包卵腺ムチンの
プロテアーゼ抵抗性断片の沈降速度図
測定温度：25℃
溶媒：0.6 M NaCl(pH 7.2)
濃度：0.045%
検出：シュリーレン光学系
写真撮影時間：最高回転数 50,000 rpm
 到達後の 13 分
沈降方向：左→右

にホモジナイザーで懸濁液とし，4℃-18 h の撹拌・抽出の後に遠心分離を行って上清液を水溶性画分とした．次に，生じた沈殿を 150 ml の 0.6 M NaCl(pH 7.2)を用いて同様に撹拌・抽出を行い，遠心分離した上清液を塩溶性画分および沈殿を不溶性画分とした．水溶性，塩溶性および不溶性の各画分は重量比で 20：24：56 であった．水溶性成分は 0.6 M NaCl に溶かしてもほとんど粘性を示さず，50%エタノールで沈殿しなかった．これに反し，塩溶性成分は 0.6 M NaCl に溶けて粘稠な溶液を生じるので，50%エタノールを加えて沈殿として回収した．

初めに，可溶性成分のアミノ酸・糖組成と SDS-ゲル電気泳動の分析を

表5 アルゼンチンイレックス包卵腺粘質物の可溶性成分の
アミノ酸・糖組成(残基数/1000)

	水溶性成分	塩溶性成分	
	全体	全体	ムチン
アミノ酸			
タウリン	173	0	0
アスパラギン酸	55	64	16
トレオニン	37	82	112
セリン	36	28	4
グルタミン酸	68	47	12
プロリン	53	54	58
グリシン	54	44	14
アラニン	55	29	7
シスチン/2	7	26	0
バリン	31	34	13
メチオニン	10	8	2
イソロイシン	25	54	56
ロイシン	37	34	10
チロシン	16	21	5
フェニルアラニン	16	20	7
トリプトファン	1	2	0
リシン	42	50	15
ヒスチジン	12	9	3
アルギニン	22	20	7
小計	750	626	341
中性糖			
ラムノース	7	21	41
フコース	46	62	91
リボース	42	14	35
キシロース	12	28	46
マンノース	53	14	8
グルコース	52	12	31
ガラクトース	21	91	168
4-O-メチルグルコース[*1]	—	—	—
小計	233	242	420
アミノ糖[*2]			
グルコサミン	10	20	36
4-O-メチルグルコサミン	3	53	80
ガラクトサミン	4	59	123
小計	17	132	239
合　計	1000	1000	1000

*1：分析を行った1994年当時,4-O-メチルグルコースの存在は不明.
*2：天然にはアミノ基がアセチル化している.

行った.表5に示すように,水溶性成分はムチンをほとんど含まず,エキス成分のタウリンが多量に認められる.50％エタノールで沈殿する塩溶性成分は未分画の粘質物と良く似たアミノ酸・糖組成をもち,トレオニン,フコース,ガラクトース,N-アセチルガラクトサミンおよびN-アセチル-4-O-メチルグルコサミンに富み,かなり多量のムチンを含むと思われた.水溶性と塩溶性の両成分はSDS-ゲル電気泳動図が明らかに異なる(図13).水溶性成分はゲル上面の高分子量成分を含まず,還元剤処理した泳動図に

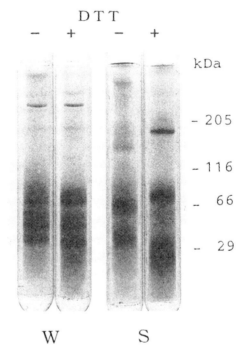

図13 アルゼンチンイレックス包卵腺粘質物の可溶性成分のSDS-ゲル電気泳動図
分析条件は図4と同じ.
W：水溶性成分
S：塩溶性成分

大きな変化がない．一方，塩溶性成分はゲル上面に留まる高分子量物質が還元剤処理で消失し，数種の新たなタンパク質が出現した．

次に，塩溶性成分の分子形と分子量を 0.6 M NaCl を含む 0.02 M リン酸ナトリウム (pH 7.2) を溶媒として調べた．図 14 に示す固有粘度 $[\eta]$ は 10 dl/g であり，この値を Simha の式に代入して得られる偏長回転楕円体の軸比は 180 である．図 15 の沈降速度図は単一の比較的鋭いピークを示し，種々の濃度で測定した沈降係数 s から濃度 0% に外挿した沈降定数 (または極限沈降係数) s^0 は 27.7 S となる．$[\eta]$ と s^0 の値を Scheraga-Mandelkern の式[5]に代入して算出した概略の分子量は著しく大きな 660 万であった．なお，$[\eta]$ と s^0 による分子量の計算法は次節の「II. スルメイカ」の項で簡潔に説明をする．

最後に，塩溶性成分からムチンの単離を試みた．希アルカリ処理で調製したムチンはコアタンパク質 34.1% および糖質 65.9% (表 5) で，固有粘

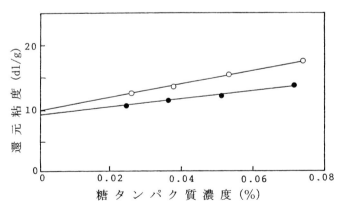

図 14 アルゼンチンイレックス包卵腺粘質物の塩溶性成分の固有粘度
測定温度：25℃，溶媒：0.6 M NaCl (pH 7.2)
○：塩溶性成分，$[\eta]$ = 10.0 dl/g
●：塩溶性成分由来のムチン，$[\eta]$ = 9.0 dl/g

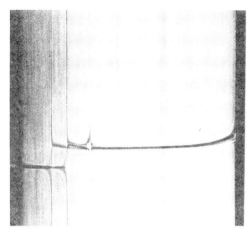

図15 アルゼンチンイレックス包卵腺粘質物の
塩溶性成分の沈降速度図
測定温度：25℃
溶媒：0.6 M NaCl(pH 7.2)
濃度：0.05%
検出：シュリーレン光学系
写真撮影時間：最高回転数 55,000 rpm
　到達後の 17 分
沈降方向：左→右

度 $[\eta]$ が 9.0 dl/g(図 14)であった．塩溶性成分の約 30％を占めるムチンと粘質物全体から分離したムチンはアミノ酸・糖組成に若干の相異があるものの，基本的には良く類似していた．

以上の結果，包卵腺粘質物の塩溶性成分は SDS-ゲル電気泳動図に認められる数種のタンパク質が会合した分子量約 660 万の巨大なムチン複合体であると考えられる．因みに，ムチンの分子量はムチン複合体の分子量とムチン含量から約 200 万と推定された．なお，未熟な包卵腺粘質物にはさらに高分子量のムチン複合体が認められ，ムチンと他のタンパク質の会合状態は包卵腺の熟度によって異なるらしい．このようなムチン複合体が第

4章で述べる卵塊膜および卵塊内部ゼリー(細繊維)の形成に関与しているのであろう.

II. スルメイカ[6]

私達がアルゼンチンイレックスの包卵腺粘質物からムチンを初めて単離したのは1994年のことであり,それから10年後の2003年に再び包卵腺ムチンをスルメイカから単離した.その理由は,1)「はじめに」で述べたように,スルメイカの産んだ貴重な卵塊を分析する機会に恵まれ,2)1997年に発表された糖分析法によって,これまで見落とされていた新規の単糖が包卵腺ムチンに発見され,後に4-O-メチルグルコースと同定されたことによる.

1. 新規の中性糖:4-O-メチルグルコース

旧来の糖分析法ではムチンの加水分解で遊離する単糖のうち,中性糖は糖アルコールのトリフルオロ酢酸(TFA)-誘導体にしてガスクロマトグラフ(蛍光法)で,一方,アミノ糖は通常のアミノ酸分析機(ニンヒドリン法)で分離定量した.中性糖とアミノ糖の加水分解条件は異なり,分析操作が煩雑であった.安野らが報告した新しい分析法は同一の加水分解(2.5 M TFA-100℃-7 h)で遊離する単糖を p-アミノ安息香酸エチルエステル(ABEE)-誘導体とし,ホーネンパックC18カラム(75×4.6 mm)を用いる逆相HPLCで中性糖とアミノ糖を同時に分離定量することができる[7].溶離液は(A):6%アセトニトリルを含む0.2 M ホウ酸カリウム緩衝液(pH 8.9)あるいは(B):10%アセトニトリルを含む0.02%トリフルオロ酢酸(TFA)を用い,溶出は45℃および流速1.0 ml/minで行った.なお,溶離液(A)による分析は加水分解物中のアミノ糖のアミノ基を N-アセチル化した後に行った.4-O-メチルグルコースの標準品は森林総合研究所の

Nishimura 博士から供与されたもの[8]で，内部標準物質は L-アラビノースを用いた．

　スルメイカ包卵腺ムチンの分析例を図 16 に示す．溶離液 B で溶出する未同定のピーク 13 は後に糖鎖の構造解析から 4-O-メチルグルコースと同定され，ガスクロマトグラフィーによる旧法ではガラクトースのピークと重なり，見落とされていたことが判明した．

2．単離と性質

　包卵腺ムチンはすでに報告した希アルカリ処理（0.4 M NaOH-4℃-4 h）で粘質物を可溶化し，0.2 M NaCl を含む 50％エタノールで生じる沈殿として調製した．粘質物の約 24％に相当する可溶化ムチンの化学組成はコアタンパク質 26.2％および糖質 73.8％で，硫酸基が単糖の 7.6％に結合している（表 6）．コアタンパク質はトレオニン，プロリンおよびイソロイシンの合計が全アミノ酸残基の約 72％を占め，これら 3 種のアミノ酸のモル比は約 2：1：1 であった．OH 基をもつセリンは著しく少なく，トレオニンがコアタンパク質と糖鎖の結合部位であるといえよう．さらに，S-S 結合に関与するシスチン/2 は検出されなかった．糖質はガラクトース，フコース，N-アセチルガラクトサミンおよび N-アセチルグルコサミンの順に多く，特に注目されるのはかなり多量の N-アセチル-4-O-メチルグルコサミン（65 残基）と 4-O-メチルグルコース（43 残基）を含んでいた．このようなメチル化糖は包卵腺ムチンに特有で，他の糖タンパク質には存在しないようである．これらの他に少量のマンノース，グルコース，キシロースおよびラムノースが検出された．

　次に 0.1 M NaCl を含む 0.02 M リン酸緩衝液（pH 7.2）を溶媒として沈降速度と粘度を 28℃で測定し，分子形と分子量を推定した．沈降速度は分析用超遠心分離機（日立モデル 282）を，粘度は速度勾配 1000 sec^{-1} の Cannon-Fenske 型毛細管粘度計を用いて測定した．なお，スルメイカの

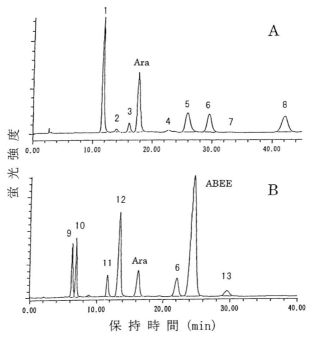

図16 スルメイカ包卵腺ムチンのABEE化単糖の分離分析
試料を加水分解($2.5\,\mathrm{M}$ TFA-100℃-7 h)の後にABEE化して，HPLCで分析した．
ABEE：p-アミノ安息香酸エチルエステル
TFA：トリフルオロ酢酸
カラム：ホーネンパックC18，45℃
流速：$1.0\,\mathrm{ml/min}$
溶離液A：6%アセトニトリルを含む$0.2\,\mathrm{M}$ホウ酸カリウム
（pH 8.9）．この系では加水分解物中のアミノ糖をN-アセチル化した．
溶離液B：10%アセトニトリルを含む0.02%TFA
ピークの同定は次のとおりである．
 1：Gal，2：Man，3：Glc，4：Xyl，5：GlcNAc
 6：Fuc，7：Rha，8：GalNAc，9：GlcN，10：GalN
 11：4-O-MeGlcN，12：Gal+Glc，13：4-O-MeGlc
 Ara：内部標準物質のアラビノース

表6 スルメイカ包卵腺ムチンの化学組成(残基数/1000)

	包卵腺ムチン	
	全 体	プロテアーゼ抵抗性断片
アミノ酸		
アスパラギン酸	9	4
トレオニン	93	96
セリン	7	8
グルタミン酸	7	3
プロリン	51	54
グリシン	8	6
アラニン	5	3
シスチン/2	2	0
バリン	9	7
メチオニン	1	0
イソロイシン	45	45
ロイシン	5	2
チロシン	2	1
フェニルアラニン	4	0
リシン	6	2
ヒスチジン	5	6
アルギニン	3	0
小計	262	237
中性糖		
ガラクトース	200	197
マンノース	7	8
グルコース	20	25
リボース	0	0
キシロース	5	9
フコース	141	176
ラムノース	7	11
4-O-メチルグルコース	43	45
小計	423	471
アミノ糖[*1]		
グルコサミン	120	116
ガラクトサミン	130	136
4-O-メチルグルコサミン	65	40
小計	315	292
合　計	1000	1000
硫酸基[*2]	56	68

*1：天然にはアミノ基がアセチル化している.
*2：硫酸基の結合している糖の残基数.

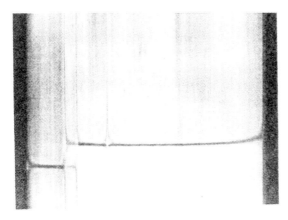

図17　スルメイカ包卵腺ムチンの沈降速度図
　　　測定温度：28℃
　　　溶媒：0.1 M NaCl(pH 7.2)
　　　濃度：0.03%
　　　検出：シュリーレン光学系
　　　写真撮影時間：最高回転数 50,000 rpm
　　　　到達後の13分
　　　沈降方向：左→右

ムチン溶液はアルゼンチンイレックスのそれと異なり，硫酸イオンによる荷電効果を消去できる NaCl の存在下で濁り（会合）を生じることなく，固有粘度と沈降定数の測定が可能であった．ムチン溶液は単一の鋭い沈降ピークを示し（図17），偏比容 v にブタ胃ムチンの文献値 0.64 ml/g [9] を用いて沈降係数 s を算出した．種々の濃度で測定した s の値の逆数を濃度に対してプロットして濃度0%に外挿した沈降定数 s^0 は 16.9 S という大きな値であった．種々の濃度で測定した還元粘度を濃度0%に外挿した固有粘度 $[\eta]$ は 12.3 dl/g である．そこで，分子形の推定に必要な容積分率固有粘度数 ν は $[\eta]$ の値を(1)式に代入し，1922 と算出された．

$$\nu = 100[\eta]/v \tag{1}$$

分子形を剛体の偏長回転楕円体と仮定して ν の値を(2)の Simha の式に代入すると, 軸比 P は 193 となり, ムチンは著しく細長い分子であることが判明した.

$$\nu = P^2/15(\ln 2P - 1.5) + P^2/5(\ln 2P - 0.5) + 14/15 \tag{2}$$

次に ν の値を Scheraga-Mandelkern の表[5]によって求めた定数 β は 3.42×10^6 であった. これらの値を(3)の Scheraga-Mandelkern の式に代入し, ムチンの分子量 M は 260 万と算出された. ただし, η_0：溶媒の粘度, N：アボガドロ数および ρ：溶媒の密度である.

$$M = [s^0[\eta]^{1/3}\eta_0 N/\beta(1-v\rho)]^{3/2} \tag{3}$$

表7　スルメイカ包卵腺ムチンとそのプロテアーゼ抵抗性断片の性質

	固有粘度(dl/g)	沈降定数(S)	分子量
ムチン	12.3	16.9	2,600,000
プロテアーゼ抵抗性断片	14.4	14.3	2,200,000

ところで, アルゼンチンイレックスの包卵腺ムチンに対するプロテアーゼの作用は先に報告した. スルメイカの包卵腺ムチンについても同様の実験を行い, 60％エタノールで沈殿するプロテアーゼ抵抗性断片は固有粘度 14.4 dl/g および沈降定数 14.3 S の各値から, 分子量 220 万と算出された (表7). すなわち, 糖鎖に比較的富む分子中央部はプロテアーゼ消化に抵抗性をもち, 分子量 220 万の巨大な断片 (アテロムチン) を生じた. 抵抗性断片はムチンの約 86％を占め, 全アミノ酸に対するトレオニン (Thr), プロリン (Pro) およびイソロイシン (Ile) の合計はムチンの 72.1％から抵抗性断片の 82.3％に増加している (表6). 因みに, アルゼンチンイレックスの

場合はムチンの 74.2% から抵抗性断片の 89.5% に増加した．さらに，Thr, Pro および Ile のモル比は約 2：1：1 なので, -Thr-Pro-Thr-Ile-, -Thr-Thr-Pro-Ile- などのアミノ酸反復配列が多く存在すると予想される．一方，抵抗性断片の単糖は 763 残基で，糖鎖がすべての Thr(96 残基)に結合すると，糖鎖は平均 8 残基の単糖を含むと推定された．

以上の結果，包卵腺ムチンは希アルカリ処理によって非共有結合が切断されて可溶化した分子量 260 万の著しく細長い棒状の巨大分子で，分子の 14% を占める末端部はトレオニンと糖鎖が比較的少なく，プロテアーゼで消化されることが判明した．

3．主要糖鎖 I の特異な構造

ムチンの糖鎖は一般に希アルカリ処理による β 離脱反応で切り出されるが，すでに述べたように，この方法は包卵腺ムチンに適用できない．そこで Kuraya & Hase が発表した温和なヒドラジン分解による糖鎖の切り出し法[10]を試みた．

ヒドラジン分解用装置ヒドラクラブ C-206(ホーネン製)を用い，ムチン 50 mg を無水ヒドラジン 15 ml で 60℃-50 h の分解を行った．分解終了の後，ヒドラジンを乾固・除去して生成する糖鎖の遊離アミノ基を N-アセチル化し，この反応液を陽イオン交換樹脂の小カラムに添加して糖鎖を水で溶出した．次に，糖鎖を Bio-Gel P4 カラム(1.8×160 cm)によるゲル濾過に付し，溶出液はオルシノール-硫酸法で中性糖を検出した．図 18 の斜線を付した主要なピーク成分を集め，糖鎖 I を単離することができた．

糖鎖 I は 1-ブタノール：酢酸：水(3：3：1, v/v)を溶媒として展開した薄層クロマトグラフ(シリカゲル-60)で単一のスポットを示す(図 19)．さらに，p-アミノ安息香酸オクチルエステル(ABOE)化した糖鎖 I は HPLC でほぼ単一成分であった．そこで，構成単糖を ABEE 化して先に述べた新しい分析法(溶離液 B)によって調べると，糖鎖 I は等モルのガラクトー

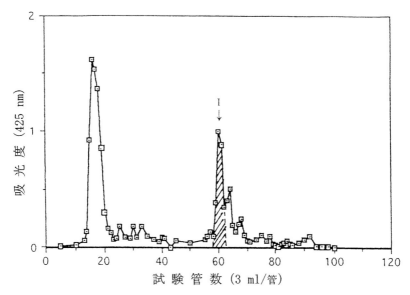

図18 スルメイカ包卵腺ムチン糖鎖のゲル濾過図
試料 50 mg を加水分解(無水ヒドラジン 15 ml-60℃-50 h)の後,糖鎖を N-アセチル化して分析した.
Ⅰ:単離した主要糖鎖
カラム:Bio-Gel P4(1.8×60 cm),55℃
流速:12.4 ml/min,溶出液:水
検出:オルシノール硫酸法

ス(Gal),フコース(Fuc),4-O-メチルグルコース(4-O-MeGlc)および 4-O-メチルグルコサミン(4-O-MeGlcN)から成っていた(図20).なお,4-O-MeGlcN は天然には N-アセチル化した 4-O-MeGlcNAc として存在している.次に,グリセロールを用いる FAB-MS で分析すると,[M+H]$^+$ に対して m/z 720 のピークが認められ,分子量は719である(図21).これらの結果,糖鎖Ⅰは Gal, Fuc, 4-O-MeGlc および 4-O-MeGlc-NAc が各1モルからなる4糖であることが判明した[6].

そこで,糖鎖Ⅰの構成糖がどのような結合状態にあるのかを解明するた

第 3 章 包卵腺のムチン　43

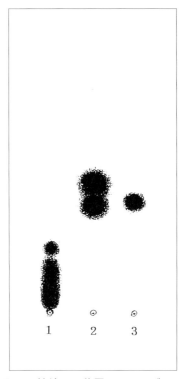

図19　糖鎖Ⅰの薄層クロマトグラム
　　　薄層板：シリカゲル-60
　　　溶媒：1-ブタノール：酢酸：
　　　　　　水(3：3：1, v/v)
　　　検出：オルシノール硫酸法
　　　　1：グルコースオリゴマー
　　　　　　(4～10糖)
　　　　2：ラフィノーズ(3糖)＋
　　　　　　ラクトース(2糖)
　　　　3：糖鎖Ⅰ

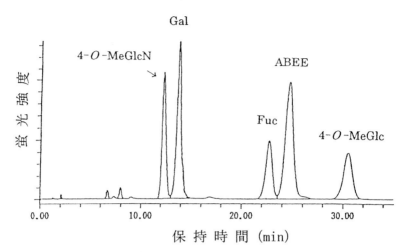

図20　糖鎖IのABEE化単糖の分離分析
図16の溶離液Bを用いて分析した.

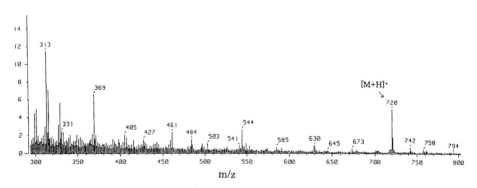

図21　糖鎖IのFAB-MSスペクトル
マトリックス：グリセロール

第3章 包卵腺のムチン　**45**

表8　糖鎖ⅠのDMSO-d_6中における^{1}Hおよび^{13}C NMRデータ

位置		^{1}H	^{13}C	HMBC相関
Gal	1	5.09(dd 3.1, 3.1)	91.9(d)	
	2	3.53(m)	79.0(d)	Gal 4, Fuc 1
	3	3.87(m)	67.7(d)	Gal 2, 4
	4	3.95(d 3.5)	78.7(d)	Gal 5, 6, 3-OH, MeGlc 1
	5	3.99(m)	67.4(d)	Gal 6
	6	3.54(dd 11.6, 5.9)	66.9(t)	Gal 5
		3.85(m)		
	1-OH	6.32(d 3.1)		
	3-OH	5.32(d 5.0)		
Fuc	1	4.73(d 3.9)	102.0(d)	Gal 2
	2	3.46(m)	68.7(d)	Fuc 3-OH
	3	3.51(m)	70.1(d)	Fuc 1, 3-OH
	4	3.45(m)	71.7(d)	Fuc 5, 6, 3-OH
	5	4.00(m)	66.4(d)	Fuc 1, 6, 4-OH
	6	1.04(d 6.7)	16.5(q)	Fuc 5
	2-OH	4.03(d 9.4)		
	3-OH	4.52(d 5.8)		
	4-OH	4.36(d 4.2)		
MeGlc	1	4.25(d 7.9)	104.5(s)	Gal 4, MeGlc 2
	2	2.98(ddd 9.0, 7.9, 2.5)	74.9(d)	MeGlc 3, 3-OH
	3	3.28(ddd 9.0, 8.6, 5.6)	76.1(d)	MeGlc 2-OH, 3-OH
	4	2.92(dd 9.8, 8.6)	79.3(d)	MeGlc 3, 4, 3-OH
	5	3.08(ddd 9.8, 5.1, 2.0)	75.6(d)	MeGlc 4, 6-OH
	6	3.46(m)	60.7(t)	MeGlc 4, 5, 6-OH
		3.60(ddd 11.8, 5.7, 2.0)		
	2-OH	5.22(d 2.7)		
	3-OH	5.03(d 5.6)		
	4-OMe	3.41(s)	59.7(q)	MeGlc 4
	6-OH	4.45(t 5.7)		
MeGlcN	1	4.35(d 8.3)	100.4(d)	Gal 6
	2	3.37(m)	55.8(d)	MeGlcN 3-OH, 3
	3	3.37(m)	74.1(d)	MeGlcN 2, 4, 5, 3-OH
	4	2.93(dd 9.8, 8.8)	79.8(d)	MeGlcN 3-OH, 4-OMe
	5	3.06(9.8, 5.1, 2.0)	75.8(d)	MeGlcN 6-OH
	6	3.48(m)	60.5(t)	MeGlcN 4, 6-OH
		3.62(ddd 11.8, 5.7, 2.0)		
	2-NH	7.64(d 9.0)		
	3-OH	5.05(d 6.0)		
	4-OMe	3.41(s)	59.6(q)	MeGlcN 4
	6-OH	4.67(t 5.7)		
	Ac-CO		169.2(s)	MeGlcN 2, 2-NH, Ac-Me
	Ac-Me	1.82(s)	23.1(q)	

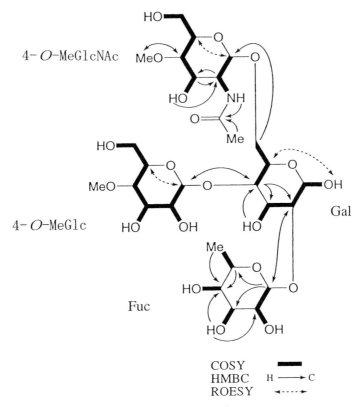

図22 糖鎖ⅠのCOSY, HMBCおよびROESY相関

め,DMSO(ジメチルスルホキシド)中で測定したNMRデータを駆使して構造解析を行った. ^{1}H および ^{13}C NMR データは表8に示す. さらに, COSY, ROESY, HOHAHA, HMBC および HBQC の各スペクトル(本章末に資料の一部を示す)による拡張二次元 NMR 分析からは図22の平面構造が決定された. これらのデータから, 糖鎖Ⅰは図23のごとく, 還元末端の Gal の C-2 に Fuc, C-4 に 4-O-MeGlc および C-6 に 4-O-MeGlcNAc の結合した分枝に富むユニークな4糖であった. 脊椎動物ムチンの糖鎖は

第3章 包卵腺のムチン 47

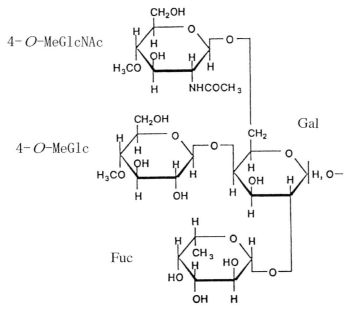

図23 糖鎖Iの4糖からなる分枝構造
Gal：ガラクトース, Fuc：フコース
4-*O*-MeGlc：4-*O*-メチルグルコース
4-*O*-MeGlcNAc：*N*-アセチル-4-*O*-メチルグルコサミン

還元末端にある GalNAc を介してコアタンパク質に結合しているのに比較すると，糖鎖Iは還元末端の Gal を介してコアタンパク質に結合している特異な構造をもつことが明らかになった．なお，この糖鎖Iの構造式は1996 年 8 〜 11 月にかけて，共同研究者の東京大学水産化学研究室の村上昌弘博士(現，共立女子大学教授)らが測定した NMR のデータから得られた結果で，その一部は 1998 年の報告書[11]に掲載した．

　一般に糖タンパク質は機能性主体がタンパク質の場合かあるいは糖鎖の場合があり，機能性主体となる糖鎖は典型的な構造とは異なる特徴を備え

ていることが多い．分泌型の包卵腺ムチンは糖含有率の高さから，糖鎖が機能性主体であることは明らかで，枝分かれに富む糖鎖Ⅰは2種のメチル化糖を含み，還元末端にガラクトースをもつなどのユニークな特徴を示すことが判明した．

【文 献】

1) Kimura, S., Sugiura, Y., Mizuno, H., Kato, N. and Hanaoka, Y.：*Fisheries Sci.*, **60**, 193（1994）
2) Bhavanandan, V. P., Sheykhnajari, M. and Devaraj, H.：*Anal. Biochem.*, **188**, 142（1990）
3) Simha, R.：*J. Phys. Chem.*, **44**, 25（1940）
4) Sugiura, Y. and Kimura, S.：*Fisheries Sci.*, **61**, 1009（1995）
5) Scheraga, H. A. and Mandelkern, L.：*J. Am. Chem. Soc.*, **75**, 179（1953）
6) Kimura, S., Gohda, T. and Sakurai, Y.：*J. Tokyo Univ. Fish.*, **89**, 7（2003）
7) Yasuno, S., Murata, T., Kokubo, K., Yamaguchi, T. and Kamei, M.：*Biosci. Biotech. Biochem.*, **61**, 1944（1997）
8) Nishimura, T., Ishihara, M., Ishii, T. and Kato, A.：*Carbohydr. Res.*, **308**, 471（1988）
9) Scawen, M. and Allen, A.：*Biochem. J.*, **163**, 363（1992）
10) Kuraya, N. and Hase, S.：*J. Biochem.*, **112**, 122（1992）
11) 木村 茂：平成8～9年度科学研究費補助金（基盤研究(B)(2)）：研究課題番号 08456103）の研究成果報告書(1998)

第3章　包卵腺のムチン　49

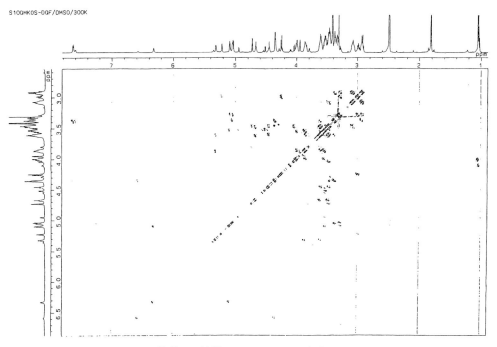

資料3　糖鎖IのCOSYスペクトル

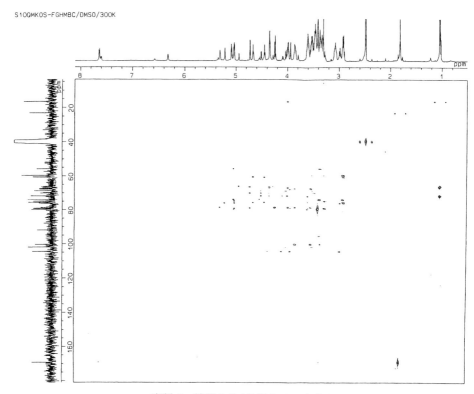

資料4 糖鎖IのHMBCスペクトル

第3章 包卵腺のムチン　51

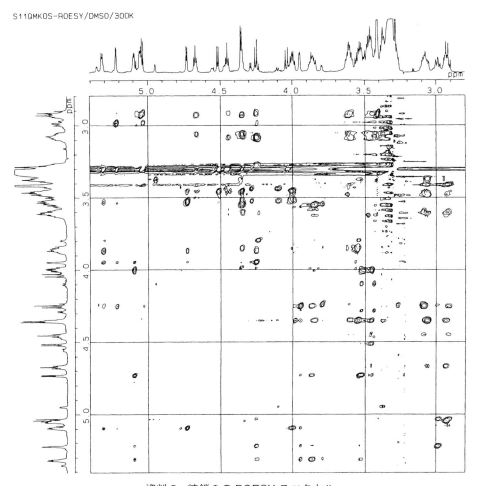

資料5　糖鎖IのROESYスペクトル

第 4 章　スルメイカの巨大卵塊[1]

　雌イカの生殖機構は種々の付属器官から構成されているが，スルメイカ類には産卵に関係する包卵腺と輸卵管腺がある．図 24 は北海道大学の水産学部付属臼尻水産実験所で観察されたスルメイカの産卵行動を示す模式図で，卵塊は表層のゲル状膜（卵塊膜）で包まれ，内部は多数の完熟卵を含む希薄なゼリー状物質（卵塊内部ゼリー）で満たされた二重構造をとっている[2]．回収された卵塊膜は両手で静かに持ち上げて広げても破れることなく，ほぼ透明なゼリー状を保ち，かなり強靭な膜構造を有している．同大学水産学部の桜井泰憲博士は 1991 年 10 月 25 日の昼に水槽で飼育しているスルメイカの産卵行動を記録しており，その一部を抜粋する[3]．

　「産卵行動は 12 時 27 分 30 秒に開始し，約 7 分で直径約 60 cm の透明な卵塊を産出した．産卵は中層で開始し，腕内部に卵塊を斜め上方から支えるように保ちながら膨らませ，最終的には雌は水槽底ぎりぎりまで沈降し，各腕をその卵塊からふりほどくようにして卵塊から離脱した．産卵前の雌は，約 1.5 m ほどの円を描くようにゆっくりと遊泳し，まず包卵腺（纏卵腺）由来のゼリー物質による小塊を全腕で小さく抱くように形成する．これに続いて，小塊内部に輸卵管腺由来の粘液状物質と卵を濾斗から連続して放出する．」

　卵塊を構成する物質の化学的研究はこれまでに無い．私達は臼尻水産実験所の水槽で飼育されているスルメイカが 1994 年 9 月 25 日に産み出した

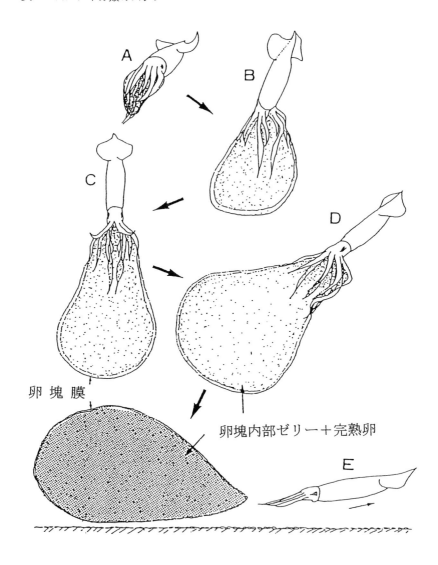

図 24　産卵行動の模式図
　　　文献 2) の図を一部改変

卵塊を桜井博士から供与していただき，包卵腺粘質物が卵塊形成に果たす役割を化学的に証明しようと試みた．卵塊膜と卵塊内部ゼリーは凍結状態で東京水産大学(現，東京海洋大学)の研究室に搬入し，低温室で解凍を行った．卵塊膜は付着する卵やゴミを除去し，水洗の後に凍結乾燥した．著しく脆弱な内部ゼリー(1000 ml)は解凍に伴って凝集した細繊維とドリップ(ゼリーの液体部分)に分離した．細繊維は付着する卵をこすり取って凍結乾燥し，一方，ドリップはガラスフィルターで濾過の後に，透析・脱塩してエバポレータで 10 ml に濃縮した．

卵塊物質との比較実験に用いる包卵腺粘質物は水溶性，塩溶性および不溶性に分画し，それらのタンパク質含量は粘質物の全タンパク質に対して各々，10, 5 および 85%であった．スルメイカの水溶性画分はアルゼンチンイレックスのそれと異なってかなり粘稠な溶液なので，等量の 6 M グアニジン塩酸を加えた後に，水溶性ムチンを 50%エタノール沈殿として単離した．不溶性画分は常法の希アルカリ処理(0.4 M NaOH-4℃-4 h)により，可溶化ムチンを 50%エタノール沈殿として単離した．

1．卵塊膜のムチン複合体

卵塊の表層ゲル状膜 1 kg(湿重量)から糖タンパク質 0.48 g が得られ，その化学組成はアミノ酸 42.7%および単糖 57.3%で，硫酸基が単糖の 11.7%に結合している(表 9)．主要成分はトレオニン，プロリン，イソロイシン，ガラクトース，フコース，N-アセチルグルコサミンおよび N-アセチルガラクトサミンで，この組成上の特徴はムチン複合体の存在を示唆している[1]．なお，卵塊膜は未同定の中性糖(X：16 残基)を含み(図 25)，4-O-メチルグルコースを欠いていた．

SDS-ゲル電気泳動で卵塊膜を分析すると(図 26)，分子量 20 万前後に 2 種およびゲル上面にいずれも不鮮明なタンパク質成分が検出され，この泳動図は DTT で S-S 結合を還元しても変化がない．それゆえ，卵塊膜はゲ

表9 卵塊膜，卵塊内部ゼリーおよび包卵腺ムチンの化学組成(残基数/1000)

	卵塊膜 全体	卵塊膜 ムチン	内部ゼリー 細繊維	包卵腺ムチン 水溶性[*1]	包卵腺ムチン 不溶性[*2]
アミノ酸					
アスパラギン酸	24	1	56	11	10
トレオニン	118	100	79	80	88
セリン	22	5	27	9	9
グルタミン酸	18	1	40	9	8
プロリン	68	54	53	45	50
グリシン	18	3	60	6	11
アラニン	10	2	26	4	6
シスチン/2	9	0	33	0	0
バリン	19	5	35	10	11
メチオニン	3	0	5	1	1
イソロイシン	62	46	53	42	48
ロイシン	12	1	27	7	5
チロシン	5	0	19	2	1
フェニルアラニン	7	0	23	3	5
リシン	19	1	49	6	6
ヒスチジン	9	5	13	5	7
アルギニン	4	0	21	2	4
小計	427	224	619	242	270
中性糖					
ガラクトース	122	156	100	173	136
X[*3]	16	25	0	23	0
マンノース	5	4	9	4	3
グルコース	30	45	6	37	15
リボース	0	0	0	0	0
キシロース	29	38	1	52	17
フコース	128	196	89	165	76
ラムノース	6	27	4	13	8
4-O-メチルグルコース	0	0	35	0	46
小計	336	491	244	467	301
アミノ糖[*4]					
グルコサミン	80	86	49	104	185
ガラクトサミン	128	161	58	164	185
4-O-メチルグルコサミン	29	38	30	23	59
小計	237	285	137	291	429
合　計	1000	1000	1000	1000	1000
硫酸基[*5]	67	60	—	56	52

*1：包卵腺粘質物の水溶性成分から調製したムチン．
*2：包卵腺粘質物から水溶性および塩溶性の成分を除いて調製したムチン．
*3：未同定の中性糖．
*4：天然にはアミノ基がアセチル化している．
*5：硫酸基の結合している糖の残基数．

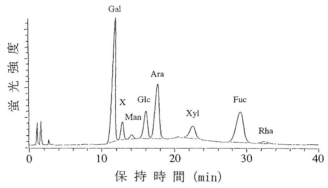

図25 卵塊膜のABEE化中性糖の分離分析
分析条件は図16の溶離液Aで行った.
Ara：内部標準物質のアラビノース
X：未同定の中性糖

ル内に泳動できないほどの高分子量のムチン複合体といえよう．一方，卵塊膜の起源と言われる包卵腺粘質物はゲル上面の高分子量成分の他に，分子量20万および4万の付近に2種の成分が鮮明なバンドとして認められ，この泳動図はDTTで還元すると明らかに変化する．新たに出現した5種ほどの成分はS-S結合をもつ非ムチン性糖タンパク質と考えられる．

卵塊膜は希アルカリ処理（0.4 M NaOH-4℃-4 h）で容易に溶けて粘稠な溶液となり，この溶液から可溶化ムチンを0.2 M NaClを含む50％エタノール沈殿（ムチン50）および75％エタノール沈殿（ムチン75）として調製した．卵塊膜ムチンの収量は重量比で卵塊膜の75％（ムチン50：50％，ムチン75：25％）であり，両ムチンはほぼ同じ化学組成をもっていた．主要なムチン50（以下，ムチンと略す）はコアタンパク質22.4％および糖質77.6％からなり，硫酸基は単糖の7.7％に結合している（表9）．コアタンパク質のトレオニン，プロリンおよびイソロイシンのモル比は約2：1：1で，これら3種のアミノ酸の合計が全アミノ酸の89％を占めていた．なお，

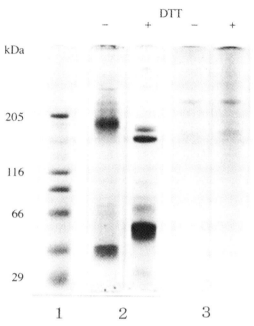

図 26　卵塊膜と包卵腺粘質物の SDS-ゲル電気泳動図
分析条件は図 4 と同じ.
1：標準タンパク質
2：包卵腺粘質物
3：卵塊膜
DTT：ジチオトレイトール（還元剤）

ムチンは包卵腺粘質物の水溶性画分から単離したムチンと良く似た化学組成をもち(表 9)，4-O-メチルグルコースを欠くが，未同定の中性糖 X(25 残基)が認められた.

次に卵塊膜ムチンの分子量を固有粘度と沈降定数の測定から推定した. 0.1 M NaCl を含む 0.02 M リン酸緩衝液(pH 7.2)に溶かしたムチンは固有粘度 $[\eta]$ が 16.6 dl/g で，沈降定数 s^0 が 16.1 S の単一の鋭いピークを示す(図 27). 分子形を剛体の偏長回転楕円体と仮定して s^0 と $[\eta]$ を

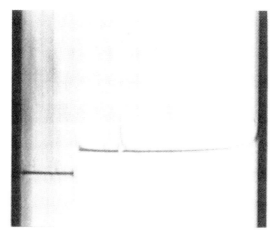

図27　卵塊膜ムチンの沈降速度図
　　　測定温度：25℃
　　　溶媒：0.1 M NaCl(pH 7.2)
　　　濃度：0.049%
　　　検出：シュリーレン光学系
　　　写真撮影時間：最高回転数 50,000 rpm
　　　　到達後の17分
　　　沈降方向：左→右

表10　スルメイカの卵塊膜ムチンと包卵腺ムチンの性質

	固有粘度(dl/g)	沈降定数(S)	分子量
卵塊膜	16.6	16.1	2,700,000
包卵腺	12.3	16.9	2,600,000

Scheraga-Mandelkernの式に代入し，分子量は270万と算出された．因みに，包卵腺粘質物と卵塊膜に由来するムチンの分子量はほぼ同一である（表10）．

2. 卵塊内部ゼリーのムチン複合体

　内部ゼリーは凍結—融解によって容易に構造が破壊され，凝集した細繊維とドリップに分かれ，ゼリー 1 L(リットル)は細繊維 15 mg とドリップ 0.1 mg のタンパク質を含んでいた．細繊維は予想図(図 28)のごとく完熟卵を囲んで卵の癒着と偏在を防ぐと考えられ，アミノ酸 61.9% と単糖 38.1% からなり，トレオニン，プロリンおよびイソロイシンに比較的富むムチン複合体といえる(表 9)．細繊維と包卵腺粘質物の不溶性成分は良く似たアミノ酸組成を示し(表 11)，中性糖組成もかなり多量の 4-O-メチルグルコースを含み，未同定の中性糖(X)を欠くなど共通の特徴が認められる(表 9)．さらに，細繊維と包卵腺粘質物を DTT 存在下で SDS-ゲル電気泳動を行うと，細繊維は包卵腺粘質物とほぼ同一の泳動図を示して輸卵管腺粘質物とは明らかに異なっており(図 29-A)，その起源は主に包卵腺粘質物の不溶性成分(図 29-B)であることが明らかになった．

　一方，ドリップのタンパク質は量的にわずかであるが，包卵腺および輸卵管腺の粘質物と異なるアミノ酸組成を示し，多量のセリン，グルタミン酸およびグリシンを含んでいた(表 11)．

図 28　卵塊内部ゼリーの細繊維(予想図)

表11 卵塊内部ゼリー，包卵腺粘質物および輸卵管腺粘質物のアミノ酸組成（残基数/1000）

アミノ酸	卵塊内部ゼリー 細繊維	卵塊内部ゼリー ドリップ	包卵腺粘質物 全体	包卵腺粘質物 不溶性[1]	輸卵管腺粘質物 全体
タウリン	0	0	26	0	21
アスパラギン酸	90	80	91	96	75
トレオニン	128	58	118	117	80
セリン	44	152	45	40	42
グルタミン酸	65	122	76	73	76
プロリン	85	37	79	80	56
グリシン	97	143	86	95	131
アラニン	42	63	43	42	44
シスチン/2	53	6	58	68	93
バリン	56	34	50	54	41
メチオニン	8	6	10	9	7
イソロイシン	85	41	73	70	44
ロイシン	43	42	36	39	20
チロシン	31	22	36	37	28
フェニルアラニン	37	23	33	39	50
X[2]	2	39	0	0	0
リシン	79	47	81	79	134
ヒスチジン	21	43	20	20	9
アルギニン	34	42	39	42	49
合 計	1000	1000	1000	1000	1000

*1：粘質物から水溶性および塩溶性の成分を除去した物質．
*2：未同定のアミノ酸．

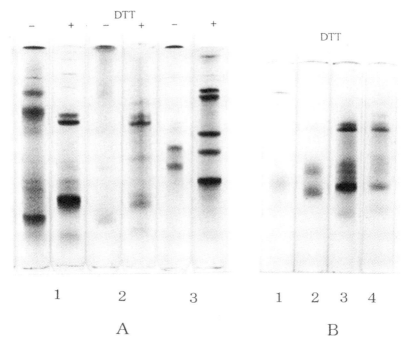

図29 卵塊内部ゼリーの細繊維と粘質物のSDS-ゲル電気泳動図
分析条件は図4と同じ.
    A. 1：包卵腺粘質物
       2：卵塊内部ゼリーの細繊維
       3：輸卵管腺粘質物
       DTT：ジチオトレイトール（還元剤）
    B. 1：包卵腺粘質物の水溶性成分
       2：包卵腺粘質物の塩溶性成分
       3：包卵腺粘質物の不溶性成分
       4：卵塊内部ゼリーの細繊維

3．卵塊形成と包卵腺粘質物

これまでの諸結果から，包卵腺粘質物は水溶性ムチン複合体が主に表層の卵塊膜を，不溶性ムチン複合体が卵塊内部ゼリーの細繊維を形成することが判明した．これら2種のムチン複合体のムチンは極めて良く似たアミノ酸組成をもち，同一のコアタンパク質に由来すると思われるが，確証はない．

次に，産卵に伴って消費される包卵腺粘質物の量はどの程度であろうか．以下のような概算を行ってみた．

① 消費された粘質物：桜井博士は1992〜3年にイカの飼育実験に際して，産卵直前のイカ3尾と産卵後の1尾および実際の産卵海域における産卵後の3尾について，外套長，包卵腺重量などを計測した（私信：2002年4月11日付の電子メール）．外套長25 cmに平均化した結果，包卵腺重量は産卵前の21.3 gおよび産卵後の12.1 gとなり，産卵に伴う粘質物の消費量は9.2 gと見積もられた．

② 卵塊膜の粘質物：卵塊は直径60 cm，卵塊膜の厚さ0.1 cm（CCDカメラによる映像から推定）および密度1.0 g/cm^{3}と仮定すると，卵塊膜の重量は1.13 kgとなる．卵塊膜1 kgは糖タンパク質0.48 gを含むので，卵塊膜1.13 kgでは0.54 gとなる．

③ 卵塊内部ゼリーの粘質物：内部ゼリー1 Lは細繊維（糖タンパク質）15 mgおよびドリップ中のタンパク質0.1 mgを含むので，直径60 cmの卵塊が含む113 Lの内部ゼリーでは1.71 gとなる．

④ 卵塊に含まれる粘質物含量：卵塊膜の0.54 gと内部ゼリーの1.71 gの合計2.25 gから，粘質物の水分76%を補正すると，卵塊に含まれる粘質物は9.4 gとなる．

以上のように，産卵に伴って消費された包卵腺粘質物の実測量（9.2 g）は直径60 cmと仮定した卵塊に含まれる粘質物の計算量（9.4 g）と不思議なほどに良く一致することが判明した．包卵腺粘質物は卵塊膜に1/4，卵

塊内部ゼリーに 3/4 が利用されており，かなりの部分が卵塊内部ゼリーの形成に関与していることが明らかになった．

ところで，「輸卵管腺粘質物は卵の正常な発生に必要な卵膜と胚の間のすき間（囲卵腔）の形成に重要である．」ことが知られている[4]．また，「輸卵管腺粘質物と卵は濾斗から卵塊内部に放出される．」といわれており，輸卵管腺粘質物は卵が円滑に体外に放出されるための潤滑剤としても役立っていることは確かであろう．しかし，輸卵管腺粘質物は包卵腺粘質物に比べて量的に著しく少なく（約 1/25），卵塊内部ゼリーの主成分とはなり難い．それゆえ，これまでの「卵塊の内部ゼリーは輸卵管腺粘質物に由来する．」という仮説は正確とはいえない．卵を包む輸卵管腺粘質物は卵塊内部ゼリーの一部分を担っているとしても，本実験によって，卵塊を実質的に支えている卵塊膜および卵塊内部ゼリーの主成分である細繊維は包卵腺粘質物由来であると結論された．今後，卵塊膜および細繊維の光学顕微鏡あるいは電子顕微鏡による観察が必要であろう．

最近，桜井博士は著書[5]の第 5 章「巨大卵塊の謎を解く」の中で，大型水槽におけるスルメイカの産卵行動，生み出された巨大卵塊の姿，回収された卵塊膜などの貴重な多数の写真を示されており，大変に興味深い．

【文 献】
1 ）Kimura, S., Higuchi, Y., Aminaka, M., Bower, J. R. and Sakurai, Y.：*J. Moll. Stud.*, **70**, 117（2004）
2 ）Sakurai,Y., Bower, J. R. and Ikeda, Y.：*Fisken og Havet.*, **12**, 105（2003）
3 ）桜井泰憲：日本海ブロック試験研究集録，28 号，1（1993）
4 ）Ikeda, Y., Sakurai, Y. and Shimazaki, K.：*Invertebrate Reproduction and Development*, **23**, 39（1993）
5 ）桜井泰憲：イカの不思議　季節の旅人・スルメイカ，北海道新聞社，pp. 1 ～ 207（2015）

第5章　包卵腺ムチンの利用[1,2]

　ムチンは唾液をつくる顎下腺，呼吸や消化の諸器官の内面を覆う粘膜などに分布し，多くの生理機能が認められている．
　① 外界からの物理的刺激に対する緩和作用，② バクテリア，ウイルス，化学汚染物質，花粉などの異物に対する接着凝集作用，③ 保湿作用，④ 潤滑作用，⑤ 酵素に対する阻止作用，⑥ 脂質に対する乳化および分散作用，⑦ 腸管の吸収強化作用，⑧ その他．
　これらの機能に着目したムチンの利用に関する研究は少ないが，例えばウシ顎下腺ムチン，ブタ胃ムチンなどは皮膚外用剤[3]や化粧品原料[4]として有用であると報告されている．しかし，原料の量的確保が難しく，粘性の高いムチンの分離精製は煩雑で，製品化された例は極めて少ない．ムチンは優れた機能特性を備えているものの，コラーゲン，ヒアルロン酸，コンドロイチン硫酸などの生体高分子物質が医薬品，医薬部外品，化粧品および健康食品に広く利用されているのとは対照的に，一般の人には馴染みが少ない．
　一方，包卵腺ムチンは原料の粘質物が比較的大量に確保でき，しかも受精卵を保護する働きは皮膚の健康と機能の保持に重要とされる要因とも一致する．さらに，私達の考案した包卵腺ムチンの単離法は効率的かつ簡便な「希アルカリ可溶化」を特徴としている．包卵腺粘質物はムチンと多くの非ムチン性タンパク質からなる複合体であり，希アルカリ処理によって非共有結合が切断されるのであろう．この単離法は包卵腺ムチンが脊椎動物ムチンと異なって希アルカリに安定で，β離脱反応によって糖鎖を失う

ことがないという奇妙な現象に気付いた成果である．希アルカリ可溶化は脊椎動物ムチンにとって常識外のことであるが，包卵腺ムチンの産業的利用を考慮した工場生産には好都合な発明といえる．このような理由から，私は東京海洋大学を定年の後，(株)高研で包卵腺ムチンの商品化と化粧品素材としての利用を試みた．

1．ムチンの製造[5]

包卵腺ムチンの小規模な工業的製法は次の2工程からなる．

① アルカリ可溶化：包卵腺粘質物をNaOH($0.1 \sim 1.0$ M)で溶解する．
② エタノール沈殿：溶解液にNaClを含むエタノール($50 \sim 65\%$)を加え，ムチンを繊維状沈殿として分離する．

主な作業の概要は次のとおりである．

　イ．包卵腺粘質物を挽肉機にかける．
　ロ．粘質物5 kgに水30 kgを加えて微粉砕する．
　ハ．NaOH処理($4℃$-4 h)によって粘質物を溶かす．
　ニ．エタノールとNaClを加えてムチンを繊維性沈殿とする．
　ホ．ムチンを水に溶かし，再度のエタノール沈殿を行う．
　ヘ．繊維状ムチンをエタノール脱水してほぐし，風乾する．

風乾したムチンの収量は包卵腺重量の$5 \sim 7\%$で，ほぼ純粋なムチンが得られることは化学組成分析の結果から確認された．さらに，このムチンを水に溶かしてプロテアーゼ消化を行うと，先に述べた透明性の高いアテロムチンを調製できる．

2．化粧品素材としての性質[2]

希アルカリ可溶化法で製造した包卵腺ムチンの0.3および1%水溶液は化粧品素材として，「マリンムチン」の商品名で2004年秋に(株)高研から発売されている．その折に，化粧品素材として重要と考えられる2, 3の

性質を調べた．以下ではマリンムチンを単にムチンと呼ぶ．

1）安全性

「化粧品の安全性評価に関する指針(2001年)：日本化粧品工業連合会編」に則り，イ) 皮膚一次刺激性試験，ロ) 累積皮膚刺激性試験，ハ) 眼刺激性試験，ニ) 皮膚感作性試験，ホ) 変異原性試験，ヘ) 急性毒性試験，ト) ヒトパッチ試験などを実施した．これらの試験の結果，問題は認められず，ムチンの安全性が確認された．なお，原料の包卵腺粘質物に関しては重金属の分析を行ったが，水銀は検出されず（検出限界 0.01 ppm），その他の重金属は微量であり，重金属汚染の心配はない．

次に，ムチンの抗原性を哺乳類アテロコラーゲン（分子末端を酵素で分解した抗原性の低い医療用コラーゲン）と比較検討した結果，ムチンはアテロコラーゲンに比べて抗原性がかなり低い(図 30)．糖タンパク質のム

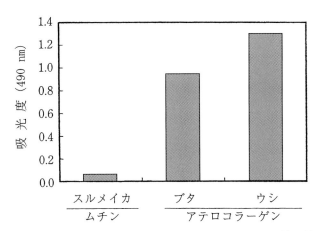

図30　ELISA 法によるムチンとアテロコラーゲンの抗原性(IgG)の比較
　　　IgG 抗体の産生量はモルモットの背部皮内に試料を投与して1週間後に追加免疫を行い，5週間後に採血して ELISA 法で測定した．

チンは糖鎖が分子の約 80%を占めるために抗原になり難いと考えられる．

2）溶液の粘性

粘性は保湿剤，潤滑剤などの評価にとって重要な因子である．そこで，ムチン，アテロムチンおよびサクシニル化したアテロコラーゲン SS（化粧水などへの配合処方に適した中性域で高い溶解性と保湿性をもつ化粧品用コラーゲン）の 0.3%水溶液（pH 6.2）を試料として，コーンプレート型回転粘度計（EHD，東京計器製）による粘度測定を行った（図 31）．得られた値をずり速度 $1\,\mathrm{sec}^{-1}$ に外挿した非ニュートン粘性係数（Pa・s：パスカル秒）はムチン：0.692 Pa・s，アテロムチン：0.396 Pa・s およびアテロコラーゲン SS：0.178 Pa・s であり，濃度 0.3%の場合，ムチンとアテロムチンはアテロコラーゲン SS より高い粘性をもっていた．ムチンは用途に応

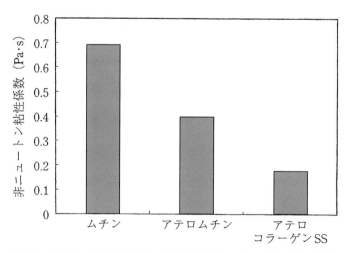

図 31 ムチン，アテロムチンおよびアテロコラーゲン SS の粘度
　　　粘度計：コーンプレート型回転粘度計
　　　試料：0.3%水溶液（pH 6.2）
　　　測定温度：20℃
　　　Pa・s：パスカル秒

じて種々の pH や NaCl 濃度で使用されるが,粘度は pH 6 〜 10 および塩濃度 0.02 〜 1 M の範囲でほぼ一定の値を示した．

3）溶液の耐熱性[6]

ムチンの 0.2％水溶液を通常の滅菌条件（オートクレーブ：121℃-1kg/cm^2-15 min）で加熱しても，粘度は見た目に大きな変化がなく，沈殿や変色も見られなかった．ムチンはこの厳しい加熱によっても SDS-ゲル電気泳動図とアミノ酸組成に変化がなく，ペプチド結合の切断は認められないが，アミノ糖の一部（約 15％）はアミノ基を失い破壊されていた．なお，加熱した溶液を 0.01％と 0.02％に希釈して測定すると，還元粘度は未加熱溶液に比べて約 1/2 に低下していた．このように，ムチンは部分的に構造破壊を生じているものの，非常に耐熱性の高いことが特徴的である．

4）凍結乾燥品の吸湿性

ムチン，アテロムチンおよびアテロコラーゲン SS の 1％溶液を水に対して透析の後，容器に一定量を入れて凍結乾燥し，相対湿度 85％の環境中に放置して経時的に重量変化を測定した．各試料は同じような吸湿曲線を示すが，ムチンとアテロムチンはアテロコラーゲン SS に比べて吸湿率がわずかに高く，保湿剤としての効果が期待される．因みに，放置 46 時間で比較すると，アテロコラーゲン SS の吸湿率は 23％で，ムチンとアテロムチンのそれは 28％であった。

以上の他に毛髪の傷みを抑える感触改善効果なども認められているが，包卵腺ムチンを化粧品素材として実際に評価することは化粧品の専門家にゆだねざるを得ない．幸いなことに，マリンムチンはわが国の大手化粧品会社が 2010 年秋に発売した最高級化粧品（コスメデコルテ AQ MW）のキー（鍵）成分に採用されている．発売から 6 年を経てようやく化粧品の専門家に保湿成分として高い評価を受けることが出来た．包卵腺ムチンは海中の受精卵を保護するという細胞に優しい物質であり，秘められた多くの機

能の解明が待たれる．今後は医薬部外品，さらに医薬品としての利用を望みたい．

【文　献】

1) 木村 茂：BIO INDUSTRY, **22**, 74 (2005)
2) 阿蘇 雄，木村 茂：*Fragrance Journal*, **34**, 14 (2006)
3) 特開 199900 (1994)
4) 山下登喜雄ほか：粧技誌，**27**, 573 (1994)
5) 特許公報(B2)P4054698 (2008)
6) 松岡玲奈：マツイカ包卵腺ムチンの耐熱性に関する研究，卒業論文，東京水産大学，東京，平成 15 年度

付　録

記載する原著論文は次の 2 編である．

1) 英国の The Malacological Society of London（ロンドン軟体動物学会）が発行する *J. Moll. Stud.* (2004) **70** : 117-121 に発表した「スルメイカの卵塊ムチン複合体の化学的性質」
2) 東京水産大学が発行する *Journal of Tokyo University of Fisheries*, **89**, 7-13 (2003) に発表した「スルメイカ包卵腺ムチンの性質」

CHEMICAL PROPERTIES OF EGG-MASS MUCIN COMPLEXES OF THE OMMASTREPHID SQUID *TODARODES PACIFICUS*

SHIGERU KIMURA[1], YUMI HIGUCHI[1], MASAKI AMINAKA[1], JOHN R. BOWER[2] AND YASUNORI SAKURAI[3]

[1]*Tokyo University of Fisheries, 4-5-7 Konan, Minato-ku, Tokyo 108-8477, Japan*
[2]*Northern Biosphere Field Science Center, Hokkaido University, 3-1-1 Minato-cho, Hakodate, Hokkaido, 041-8611, Japan*
[3]*Graduate School of Fisheries Sciences, Hokkaido University, 3-1-1 Minato-cho, Hakodate, Hokkaido 041-8611, Japan*

(Received 17 February 2003; accepted 5 August 2003)

ABSTRACT

This paper describes the properties of egg-mass glycoproteins from the Japanese common squid *Todarodes pacificus* (Steenstrup, 1880) and provides chemical evidence for the role of a nidamental mucosubstance in the formation of egg masses. Samples were collected from the surface layer and the interior jelly of an egg mass spawned in captivity, and the interior jelly was separated into fibrils and the liquid portion of the jelly. Mucosubstances in the nidamental and oviducal glands of maturing squid were then isolated and compared with those of each part of the egg mass. The surface layer and the interior fibrils were found to be composed of distinct types of mucin complexes and to have originated from the water-soluble and insoluble fractions of nidamental mucosubstance, respectively. Oviducal mucosubstance, however, appeared to play no role in the formation of the fibrils. These results suggest the egg-mass surface layer is derived from the water-soluble fraction of the nidamental mucosubstance, and the fibrils within the mass are derived from the insoluble fraction of the mucosubstance.

INTRODUCTION

The female reproductive system of squids comprises several accessory organs, which generally include the nidamental gland and oviducal gland (Nesis, 1987; Budelmann, Schipp & Boletzky, 1997). Both glands secrete substances when egg masses are spawned; the oviducal gland secretes a viscous substance that envelops the eggs, and the nidamental gland secretes an albuminous substance that forms the egg-mass surface layer (Hamabe, 1962).

Like other ommastrephid squids, the Japanese common squid *Todarodes pacificus* (Steenstrup, 1880) spawns spherical, gelatinous egg masses (Bower & Sakurai, 1996). The egg masses can reach at least 80 cm in diameter and are composed of two parts: a jelly-like surface layer and the egg-mass interior, which contains eggs surrounded by jelly (Hamabe, 1961, 1963; Bower & Sakurai, 1996). To date, there is no information on the chemical properties of either the egg-mass surface layer or interior. The present paper describes the properties of egg-mass glycoproteins from *Todarodes pacificus* and provides chemical evidence for the role of a nidamental mucosubstance in the formation of egg masses.

MATERIAL AND METHODS

Egg-mass analysis

A *Todarodes pacificus* egg mass spawned in captivity on 25 September 1994 at the Usujiri Fisheries Laboratory (Hokkaido University, Japan) was analysed. Samples were collected from the surface layer and the interior jelly, and stored at −30°C until analysis. The surface-layer sample was thawed in a cold room, washed with water to remove any attached eggs or contaminants, and lyophilized. A total of 1000 ml of the frozen interior jelly was separated into fibrils and drip (the liquid portion of the jelly) by thawing. The fibrils were cleaned by scraping off any attached eggs, washed with water, and lyophilized. The drip was filtered through a glass filter, dialysed against water and concentrated using a rotary evaporator to 10 ml.

Preparation of nidamental and oviducal mucosubstances

To analyse the mucosubstances in nidamental and oviducal glands of *Todarodes pacificus*, samples of both glands were removed from maturing squid collected in inshore waters of Tsugaru Strait, southern Hokkaido, Japan. The mucosubstances of each gland were isolated after removing the thin, collagenous membranes surrounding the glands.

The nidamental mucosubstance was separated into three fractions based on solubility: water-soluble, salt-soluble and insoluble. A total of 2 g of the mucosubstance was homogenized with 80 ml of water and stirred at 4°C for 1 day. The homogenate was then centrifuged at 135,600 g for 1 h, and the supernatant was obtained as the water-soluble fraction. Next, the precipitate was homogenized with 50 ml of 3% (w/v) NaCl and stirred at 4°C for 5 days. The salt-soluble fraction was then obtained by centrifugation as described above. The residue was recovered as the insoluble fraction, which was washed thoroughly with water and lyophilized.

Preparation of mucins

Water-soluble mucin from the nidamental mucosubstance was prepared as follows. To the water-soluble fraction, guanidium chloride was added to a final concentration of 6 M. After gentle stirring of the solution for 3 h at 20°C, the mucin was precipitated with 60% ethanol containing 0.2 M NaCl and dissolved in 30 ml of water. The mucin solution was treated once more with guanidium chloride, and after precipitation and dissolution of the mucin, it was dialysed against water and lyophilized.

Insoluble mucin of the nidamental mucosubstance was prepared using the method described by Kimura, Gohda & Sakurai (2003) for alkali-solubilized mucin. To the insoluble fraction, 4 M NaOH was added to a final concentration of 0.4 M

Correspondence: S. Kimura; e-mail: kimus@tokyo-u-fish.ac.jp

to solubilize the mucin. After gentle stirring at 4°C for 4 h, the solubilized mucin was clarified by centrifugation at 90,000 g for 1 h and precipitated with 60% ethanol containing 0.2 M NaCl. The precipitate was then dissolved in 30 ml of water. After the precipitation and dissolution of the mucin was carried out once more, the mucin solution was dialysed against water and lyophilized.

Mucin from the egg-mass surface layer was prepared similarly to the method mentioned above. Briefly, 95 mg of the surface layer was homogenized with 100 ml of 0.4 M NaOH and stirred at 4°C for 3 days. Most of the surface layer was solubilized to give a highly viscous solution, and after filtration, the solubilized mucin was precipitated with 50% ethanol containing 0.2 M NaCl (mucin 50), followed by precipitation with 75% ethanol containing 0.2 M NaCl (mucin 75). The precipitates were collected by centrifugation, and each precipitate was dissolved in 50 ml of water. The precipitation and dissolution of the mucin was then repeated, and the mucin solution was clarified by centrifugation at 90,000 g for 1 h. Finally, the mucin solution was dialysed against water and lyophilized.

Analytical methods

Protein assays of samples were performed with a Micro BCA Protein Assay Reagent Kit (Pierce Biotechnology, Illinois, USA) using bovine serum albumin as a standard. Amino acids and sugars were determined by high performance liquid chromatography (HPLC) as reported previously (Kimura et al., 2003). Samples were also analysed by sodium dodecyl sulphate–polyacrylamide gel electrophoresis (SDS–PAGE) using 3.5% polyacrylamide gels in 0.1 M sodium phosphate buffer, pH 7.2, containing 0.1% SDS and 3.5 M urea at 8 mA per tube for 4 h (Kimura et al., 1994). Protein bands were stained with Coomassie Brilliant Blue R-250 dye. Ester sulphate analysis, sedimentation velocity analysis and viscosity measurement were carried out according to the method described in Kimura et al. (2003).

RESULTS

Egg-mass surface layer

The egg-mass surface layer comprised approximately 0.048% glycoproteins, plus a number of electrolytes; the glycoproteins comprised 66.2% sugar, 29.5% protein, and 4.3% ester sulphate by weight. The surface layer was abundant [>50 residues/1000 total amino acid and sugar residues (TAASR)] in threonine, proline, isoleucine, galactose, fucose, N-acetylglucosamine, N-acetylgalactosamine and ester sulphate (Table 1). This composition is sufficient to account for the occurrence in the surface layer of a sulphated mucin complex (Sugiura & Kimura, 1995), which is characterized by a high sugar:protein ratio. The surface layer had a considerable amount (16 residues/1000 TAASR) of the unidentified sugar X (Fig. 1) and no 4-O-methylglucose (Table 1).

When examined by SDS–PAGE (Fig. 2), the protein components of the surface layer gave only a few, faint bands, and no significant change was observed after reduction with dithiothreitol (DTT); the surface layer was hardly stained by Coomassie Brilliant Blue R-250 dye owing to the large number of sugar chains present in these protein components. Both electrophoretic patterns showed that the surface layer contained a mucin complex that did not migrate into the polyacrylamide gel because of its extremely high molecular mass (described below). On the other hand, the nidamental mucosubstance gave several clear bands, which changed after reduction with DTT (Fig. 2). These protein bands were probably derived from the large amounts of non-mucinous proteins in the mucosubstance.

Surface-layer mucin was easily isolated by solubilization with 0.4 M NaOH, followed by precipitation with 50% and 75% ethanol; the yield of mucin was about 75% (50% for mucin 50 and 25% for mucin 75) of the surface layer by weight. Both mucin preparations shared the same compositional features, but some significant differences were evident; for instance, the mucin 75 was relatively rich in acidic amino acids when compared with the mucin 50. The major mucin 50 was characteristic in many respects. It was made up of only 20.0% protein by weight (Table 1), and approximately 89% of the protein backbone by residues was accounted for by three amino acids,

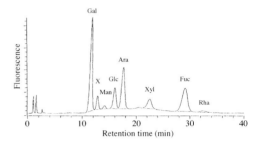

Figure 1. HPLC of p-aminobenzoic ethyl ester (ABEE)-derivatized monosaccharides from the surface layer of a Japanese common squid (*Todarodes pacificus*) egg mass. The ABEE-derivatized sugars were eluted from a Honenpak C18 column at 45°C with 0.02% trifluoroacetic acid containing 10% acetonitrile at a flow rate of 1.0 ml/min. Arabinose (Ara) was used as an internal standard. X is an unidentified sugar. See Table 1 for abbreviations.

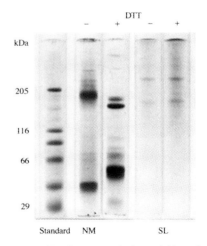

Figure 2. SDS–PAGE of the egg-mass surface layer and nidamental mucosubstance of the Japanese common squid (*Todarodes pacificus*). Samples in the presence (+) and absence (−) of dithiothreitol (DTT) were resolved by the method described in the text and visualized with Coomassie Brilliant Blue R-250 dye. Molecular mass markers (Sigma) for SDS–PAGE were used as standard proteins. Abbreviations: NM, nidamental mucosubstance; SL, egg-mass surface layer.

threonine, proline and isoleucine, which were present in a molar ratio of about 2:1:1 (Table 1). The surface-layer mucin lacked 4-O-methylglucose but contained considerable amounts of the unknown sugar X.

The molecular mass and molecular shape of mucin 50 were estimated by both sedimentation velocity analysis and viscosity measurement. A single, sharp peak having a sedimentation constant of 16.1 S was observed for the mucin solution at 25°C (Fig. 3). At the same temperature, it had a high intrinsic viscosity of $[\eta] = 16.6 \, dl \, g^{-1}$, which is indicative of a long, rod-like structure. From these results, we calculated the molecular mass of the mucin to be 2700 kDa using the Scheraga-Mandelkern equation (Scheraga & Mandelkern, 1953). The macromolecular properties of the mucins of the egg-mass surface layer and the nidamental mucosubstance were almost identical (Table 2).

Egg-mass interior jelly

The interior jelly was composed of fibrils and drip. When 1000 ml of the frozen jelly was thawed, approximately 15 mg of fibrils was recovered, and the weights of proteins in the fibrils and drip were 7.5 mg and 0.1 mg, respectively. The amino acid composition of the fibrils and the drip is listed in Table 3,

Table 1. Chemical compositions of the egg-mass glycoproteins and nidamental mucins of the Japanese common squid, *Todarodes pacificus*.

	Surface layer		Interior jelly	Nidamental mucin	
	Whole	Mucin 50*	Fibrils	Water-soluble	Insoluble
Amino acid					
Asp	24	1	56	11	10
Thr	118	100	79	80	88
Ser	22	5	27	9	9
Glu	18	1	40	9	8
Pro	68	54	53	45	50
Gly	18	3	60	6	11
Ala	10	2	26	4	6
Cys	9	0	33	0	0
Val	19	5	35	10	11
Met	3	0	5	1	1
Ile	62	46	53	42	48
Leu	12	1	27	7	5
Tyr	5	0	19	2	1
Phe	7	0	23	3	5
His	9	5	13	5	7
Lys	19	1	49	6	6
Arg	4	0	21	2	4
Sub-total	427	224	619	242	270
Neutral sugar					
Gal	122	156	100	173	136
X†	16	25	0	23	0
Man	5	4	9	4	3
Glc	30	45	6	37	15
Xyl	29	38	1	52	17
Fuc	128	196	89	165	76
Rha	6	27	4	13	8
4-O-MeGlc	0	0	35	0	46
Sub-total	336	491	244	467	301
Amino sugar					
GlcNAc	80	86	49	104	185
GalNAc	128	161	58	164	185
4-O-MeGlcNAc	29	38	30	23	59
Sub-total	237	285	137	291	429
Total	1000	1000	1000	1000	1000
Ester sulphate	67	60	–‡	56	52

Values are given in residues/1000 total amino acid and sugar residues (TAASR). *Mucin precipitated with 50% ethanol containing 0.2 M NaCl. †An unidentified component, X, was determined to be mannose equivalent. ‡Not determined. Abbreviations: Asp, aspartic acid; Thr, threonine; Ser, serine; Glu, glutamic acid; Pro, proline; Gly, glycine; Ala, alanine; Cys, cystine; Val, valine; Met, methionine; Ile, isoleucine; Leu, leucine; Tyr, tyrosine; Phe, phenylalanine; His, histidine; Lys, lysine; Arg, arginine; Gal, galactose; Man, mannose; Glc, glucose; Xyl, xylose; Fuc, fucose; Rha, rhamnose; 4-O-MeGlc, 4-O-methylglucose; GlcNAc, N-acetylglucosamine; GalNAc, N-acetylgalactosamine; 4-O-MeGlcNAc, N-acetyl-4-O-methyl-glucosamine.

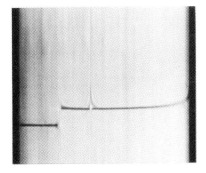

Figure 3. Sedimentation velocity pattern of mucin 50 derived from the egg-mass surface layer at 25°C. The mucin was dissolved in 0.02 M sodium phosphate buffer (pH 7.2) containing 0.1 M NaCl at a concentration of 0.049%. The photograph was taken after centrifugation for 17 min at 50,000 rpm (= 200,000 g).

Table 2. Macromolecular properties of mucins from the egg-mass surface layer and nidamental mucosubstance of the Japanese common squid.

Mucin*	$[\eta]$(dl g^{-1})	$S^0_{20,w}$ (S)	Molecular mass (kDa)
Egg-mass surface layer	16.6	16.1	2700
Nidamental mucosubstance[†]	12.3	16.9	2600

*Alkali-solubilized mucin. [†]Kimura et al. (2003).

together with those of the nidamental mucosubstance and its fractions, and of the oviducal mucosubstance. The fibrils were similar to the nidamental mucosubstance (whole) in having relatively high contents of threonine, proline and isoleucine, but were not similar to the oviducal mucosubstance. The drip differed from the fibrils in having large amounts of serine, glutamic acid and glycine.

The fibrils and the nidamental and oviducal mucosubstances were subjected to SDS–PAGE. In the presence of DTT, the band patterns of fibrils were almost identical to those of the nidamental mucosubstance but distinct from those of the oviducal mucosubstance (Fig. 4), suggesting that the fibrils were derived from the nidamental mucosubstance. Moreover, the major component of the fibrils in the absence of DTT was located at the top of the gel, indicating that it was highly polymerized with disulphide bonds.

To verify the origin of the fibrils, the three fractions of nidamental mucosubstance were examined. The relative amounts of protein in the water-soluble, salt-soluble and insoluble fractions were 10%, 5% and 85%, respectively. SDS–PAGE of these fractions clearly shows that the insoluble fraction was very similar to the fibrils, but the water-soluble and salt-soluble fractions were not (Fig. 5). This was also confirmed by the finding that the amino-acid composition of the insoluble fraction was similar to that of the fibrils (Table 3). Of these fractions, the water-soluble fraction was very viscous and typical of mucinous proteins in having relatively high amounts of threonine, proline and isoleucine. Like the surface-layer mucin, the water-soluble mucin lacked 4-O-methylglucose but contained the sugar X (23 residues/1000 TAASR) (Table 1). On the other hand, the insoluble mucin lacked X but contained 4-O-methylglucose (46 residues/1000 TAASR), closely resembling the fibrils.

Table 3. Amino acid compositions of the egg-mass interior jelly (EMIJ), nidamental mucosubstance (NM) and oviducal mucosubstance (OM) of the Japanese common squid.

	EMIJ		NM				OM
	Fibrils	Drip	Whole	Water soluble	Salt soluble	Insoluble	Whole
Tau	0	0	26	–*	0	0	21
Asp	90	80	91	81	83	96	75
Thr	128	58	118	192	144	117	80
Ser	44	152	45	78	45	40	42
Glu	65	122	76	88	79	73	76
Pro	85	37	79	118	99	80	56
Gly	97	143	86	56	88	95	131
Ala	42	63	43	53	39	42	44
Cys	53	6	58	11	49	68	93
Val	56	34	50	39	43	54	41
Met	8	6	10	9	9	9	7
Ile	85	41	73	103	86	70	44
Leu	43	42	36	38	31	39	20
Tyr	31	22	36	14	31	37	28
Phe	37	23	33	19	27	39	50
His	21	43	20	23	21	20	9
Y[†]	2	39	0	3	1	0	0
Lys	79	47	81	59	90	79	134
Arg	34	42	39	16	35	42	49
Total	1000	1000	1000	1000	1000	1000	1000

Values are given in residues/1000 total amino acid residues. *A large amount of Tau (taurine) was concentrated in the water-soluble fraction, but was omitted from this table because it was a non-proteinous amino sulphonic acid. [†]An unidentified amino acid, Y, was found between His and Lys and determined to be Lys equivalent. See Table 1 for abbreviations.

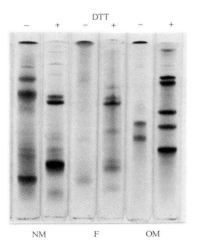

Figure 4. SDS–PAGE of the egg-mass fibrils and mucosubstances from the nidamental and oviducal glands of the Japanese common squid (*Todarodes pacificus*). The electrophoretic conditions were the same as described in Figure 2. Abbreviations: NM, nidamental mucosubstance; F, fibrils; OM, oviducal mucosubstance.

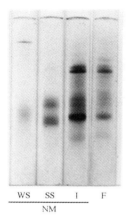

Figure 5. SDS–PAGE of the egg-mass fibrils and fractions of nidamental mucosubstance of the Japanese common squid (*Todarodes pacificus*). Samples were resolved in the presence of DTT. The electrophoretic conditions were the same as described in Figure 2. Abbreviations: NM, nidamental mucosubstance; WS, water-soluble fraction; SS, salt-soluble fraction; I, insoluble fraction; F, fibrils.

DISCUSSION

Our results show that nidamental mucosubstance plays two important roles in the formation of *Todarodes pacificus* egg masses. First, the water-soluble fraction of the mucosubstance forms the egg-mass surface layer, thus confirming Hamabe's (1962) observation that the nidamental-gland secretion forms the surface layer. Second, the insoluble fraction of the mucosubstance forms the fibrils within the egg mass. The oviducal mucosubstance, which is essential for chorion expansion in fertilized eggs (Ikeda, Sakurai & Shimazaki, 1993), surrounds each egg when they are spawned and, thus, is assumed to occur in the drip within the interior of the mass, but plays no role in the formation of the fibrils. Further histochemical studies of the interior jelly are needed to determine the role of the fibrils and confirm the source of the interior jelly.

ACKNOWLEDGEMENTS

We thank the staff of the Usujiri Fisheries Laboratory for their help rearing the pre-spawning squid and the two anonymous reviewers of the manuscript.

REFERENCES

BOWER, J.R. & SAKURAI, Y. 1996. Laboratory observations on *Todarodes pacificus* (Cephalopoda: Ommastrephidae) egg masses. *American Malacological Bulletin*, 13: 65–71.

BUDELMANN, B.U., SCHIPP, R. & BOLETZKY, S.V. 1997. Cephalopoda. In: *Microscopic anatomy of invertebrates*, 6A (Mollusca II) (F.W. Harrison & A.J. Kohn, eds), 119–414. Wiley-Liss, New York.

HAMABE, M. 1961. Experimental studies on breeding habit and development of the squid, *Ommastrephes sloani pacificus* Steenstrup. 2. Spawning behavior. *Zoological Magazine*, 70: 385–394. [Japanese with English summary]

HAMABE, M. 1962. Embryological studies on the common squid, *Ommastrephes sloani pacificus* Steenstrup, in the southwestern waters of the Sea of Japan. *Bulletin of the Japan Sea Regional Fisheries Research Laboratory*, 10: 1–45. [Japanese with English summary]

HAMABE, M. 1963. Spawning experiments on the common squid, *Ommastrephes sloani pacificus* Steenstrup, in an indoor aquarium. *Bulletin of the Japanese Society of Scientific Fisheries*, 29: 930–934. [Japanese with English summary]

IKEDA, Y., SAKURAI, Y. & SHIMAZAKI, K. 1993. Fertilizing capacity of squid (*Todarodes pacificus*) spermatozoa collected from various sperm storage sites, with special reference to the role of gelatinous substance from oviducal gland in fertilization and embryonic development. *Invertebrate Reproduction and Development*, 23: 39–44.

KIMURA, S., GOHDA, T. & SAKURAI, Y. 2003. Characterization of nidamental mucin from Japanese common squid *Todarodes pacificus*. *Journal of the Tokyo University of Fisheries*, 89: 7–13.

KIMURA, S., SUGIURA, Y., MIZUNO, H., KATO, N. & HANAOKA, Y. 1994. Occurrence of a mucin-type glycoprotein in nidamental gland mucosubstance from squid *Illex argentinus*. *Fisheries Science*, 60: 193–197.

NESIS, K.N. 1987. *Cephalopods of the world*. T.F.H. Publications, Neptune City, New Jersey.

SCHERAGA, H.A. & MANDELKERN, L. 1953. Consideration of the hydrodynamic properties of proteins. *Journal of the American Chemical Society*, 75: 179–184.

SUGIURA, Y. & KIMURA, S. 1995. Nidamental gland mucosubstance from the squid *Illex argentinus*: salt-soluble component as a mucin complex. *Fisheries Science*, 61: 1009–1011.

Characterization of Nidamental Mucin from Japanese Common Squid *Todarodes pacificus*

KIMURA Shigeru[*1], GOHDA Tomoko[*1] and SAKURAI Yasunori[*2]

(Received August 21, 2002)

Abstract: The nidamental mucosubstance, 3 g by wet weight, of Japanese common squid *Todarodes pacificus* was solubilized with 0.4 M NaOH containing 0.1 M NaBH$_4$ at 4°C for 4 h. A water-soluble mucin, 170 mg by dry weight, was isolated by precipitation with 50 % ethanol. The nidamental mucin comprised 16.4 % protein, 80.3 % sugar and 3.3 % ester sulfate by weight. About 72 % of the protein backbone was composed of threonine, proline and isoleucine in the molar ratio of about 2:1:1. The compositional feature of sugars was the presence of two methylated monosaccharides, 4-*O*-methylglucose and *N*-acetyl-4-*O*-methylglucosamine. The mucin had a long, rod-like structure with a molecular mass of 2,600 kDa and gave a very viscous solution in water. Before and after protease digestion of the mucin, no marked change of its molecular structure was observed. Sugar chains of the mucin were resistant to β-elimination, but were released by mild hydrazinolysis. A major component of the sugar chains was found to be a tetrasaccharide consisting of 1 mol each of galactose, fucose, 4-*O*-methylglucose and *N*-acetyl-4-*O*-methyl-glucosamine. These results suggest that the mucin plays a crucial role in the high viscosity and gel-forming properties of the nidamental mucosubstance.

Key words: mucin, nidamental gland, squid, Todarodes pacificus, N-acetyl-4-O-methylglucosamine, 4-O-methylglucose

Introduction

The nidamental gland mucosubstance of ommastrephid squid is assumed to form the jelly-like surface layer of egg masses up to 80 cm in diameter.[1,2] This surface layer is effective in preventing bacteria, protozoans, and crustaceans present in seawater from infesting the egg mass. We previously isolated a sulfated glycoprotein, mucin, from the nidamental mucosubstance of Argentine shortfin squid *Illex argentinus*.[3,4] Its protein backbone, which accounted for 18.7 % by weight, was characterized by extremely high levels of threonine, proline and isoleucine and was postulated to link *O*-glycosidically to sugar chains through threonine residues. An alkaline treatment of the mucin with 0.5 M NaOH at 25 °C for 24 h, however, failed to release *O*-glycosidically linked oligosaccharides, indicating that β-elimination did not occur. In addition, an unusual amino sugar was found in acid hydrolysates of the mucin.[3] and determined its chemical structure to be 4-*O*-methyl-D-glucosamine.[5] This unique mucin seems to be widely distributed in ommastrephid squid species, because they produce such large egg masses in seawater.[1,2]

One of the ommastrephid squid, Japanese common squid *Todarodes pacificus,* is a commercially important resource in Japan and two egg masses spawned by two captive *T. pacificus* females were obtained at the Usujiri Fisheries Laboratory of Hokkaido University.[2] As the first step of our biochemical study on egg mass mucin, we report here characterization of the nidamental mucin of *T. pacificus,* a possible precursor of the egg mass mucin.

Materials and Methods

Isolation of Alkali-solubilized Mucin from the Nidamental Mucosubstance

The nidamental glands of *T. pacificus* were obtained at the Usujiri Fisheries Laboratory, Hokkaido University and were stored at -30 °C until use. Nidamental mucin was isolated as a water-soluble form according to the method described previously.[3] Briefly, the gum-like mucosubstance of nidamental glands, 3 g by wet weight, was homogenized in 150 mL of distilled water and solubilized with 0.4 M NaOH containing 0.1 M NaBH$_4$ for 4 h at 4°C while actively stirring. The resulting viscous solution was clarified by centrifugation and treated with 50 % ethanol in the presence of 0.2 M NaCl to precipitate an alkali-solubilized mucin. The fibrous precipitate was collected by

---

[*1] Department of Food Science and Technology, Tokyo University of Fisheries, Minato-ku, Tokyo 108-8477, Japan.
[*2] Laboratory of Marine Ecology, Graduate School of Fisheries Sciences, Hokkaido University, Hakodate, Hokkaido 041-8611, Japan.

a pincette, washed with 60 % ethanol in the presence of 0.2M NaCl and dissolved in water. This precipitation and solution was repeated twice. The mucin solution thus obtained was dialyzed extensively against water and lyophilized.

Preparation of a Protease-resistant Fragment from Mucin

Nidamental mucin, 100 mg by dry weight, was dissolved in 100 mL of 0.5 M sodium acetate buffer (pH 8.0) containing 10 mM $CaCl_2$ and 3 % ethanol and was digested with 2 mg of actinase E from *Streptomyces griseus* (EC 3.4.24.4, Kaken Seiyaku Co.) for 17 h at 50 ℃. The protease digestion was stopped by addition of NaOH at 4 ℃ to a final concentration of 0.1 M, followed by incubation for 10 min. A protease-resistant fragment of the mucin was recovered by precipitation with 60 % ethanol in the presence of 0.2 M NaCl, dissolved in water and lyophilized.

Analytical Methods

Amino acid compositions, after hydrolysis of samples with 6 M HCl at 110 ℃ for 24 h, were determined by high-performance liquid chromatography (HPLC) with a TSK gel Aminopak column (Tosoh). Neutral and amino sugars, after hydrolysis of samples with 2.5 M trifluoroacetic acid (TFA) at 100 ℃ for 7 h,[6] were analyzed by the method of Yasuno et al.[7] Derivatization of monosaccharides was performed with a *p*-aminobenzoic ethyl ester (ABEE)-derivatized kit (Honen Corp.). ABEE-derivatized monosaccharides were subjected to reversed-phase HPLC on a 4.5 mm x 75 mm column of Honenpak C18 and were eluted with 0.02 % TFA containing 10 % acetonitrile or 0.2 M potassium borate buffer (pH 8.9) containing 6 % acetonitrile at a flow rate of 1.0 mL/min at 45 ℃. The effluent was monitored by a Tosoh FS-8010 fluorescence spectrophotometer with an excitation wavelength of 305 nm and an emission wavelength of 360 nm. L-Arabinose was used as an internal standard. The authentic compound of 4-*O*-methylglucosamine was obtained in our previous study[4] and that of 4-*O*-methylglucose was a gift from Dr Nishimura T. (Forestry and Forest Products Research Institute, Tsukuba).[8]

Determination of *N*-acetylhexosamine-terminating *O*-glycosidically linked sugar chains in the mucin was carried out with a modified Morgan-Elson reaction as reported by Bhavanandan et al.[9]

Moreover, the *N*-acetyl group of amino sugars was assayed by HPLC on a TSK gel OApak column (Tosoh) as acetic acid liberated by hydrolysis with 2 M HCl at 100 ℃ for 2 h[3] and ester sulfate was by the sodium rhodizonated method.[10]

Sedimentation velocity analysis was performed with an analytical ultracentrifuge (Hitachi model 282) at 50,000 rpm and at 28 ℃. The values of sedimentation coefficients were calculated by using a partial specific volume of 0.64 mg/mL for mucin.[11] Viscosity measurement was carried out at 28 ℃ with a Cannon-Fenske viscometer having an average shear gradient of about 1,000 sec^{-1}. Samples were dissolved in 0.02 M sodium phosphate buffer (pH 7.2) containing 0.1 M NaCl.

Fast atom bombardment (FAB) mass spectrometry was performed in the positive ion mode on a JEOL JMS SX-102 mass spectrometer.

Preparation of O-linked Sugar Chains from Mucin

O-linked sugar chains of the nidamental mucin were released by mild hydrazinolysis. In order to avoid excessive degradation of the sugar chains, the mucin (50 mg) was treated with 1.5 mL of anhydrous hydrazine at 60℃ for 50 h as reported by Kuraya and Hase[12] using a hydrazinolysis instrument "Hydraclub C-206" (Honen Corp.). After removal of hydrazine by repeated evaporation *in vacuo*, free amino groups of the released sugar chains were *N*-acetylated with acetic anhydride and saturated sodium bicarbonate solution. The reaction mixture was left in an ice bath for 15 min with occasional stirring and poured into a small column of Dowex 50W-X8 (H^+) resin. The column was washed with 5 bed volumes of water, and the passed-through fraction and the washings were combined and lyophilized.

Sugar chains released by hydrazinolysis of the mucin were subjected to gel filtration on a 1.8 cm x 160 cm column of Bio-Gel P4 (Pharmacia) and eluted with water at a flow rate of 12.4 mL/h at 55 ℃. The effluent was monitored by the orcinol-sulfuric acid method[13] and appropriate fractions were collected and lyophilized. A major sugar chain was rechromatographed on the same Bio-Gel P4 column, derivatized with a *p*-aminobenzoic octyl ester (ABOE)-derivatized kit (Honen Corp.) and analyzed by reversed-phase HPLC on a 4.6 mm x 75 mm column of Honenpak C18. Chromatography was carried out at 45 ℃ with two solvents, A and B, as the eluent at a flow rate of 1.0 mL/min ; solvent A was a 75:25 mixture of water and acetonitrile and solvent B was a 55:45 mixture. The column was equilibrated with solvent A. After injection of a sample, the ratio of solvent B was 0 % for 10 min, then it was increased by a linear gradient of 10 -100

% in 60 min. The effluent was monitored with the fluorescence spectrophotometer as mentioned above.

Results and Discussion

Properties of Nidamental Mucin

From the nidamental mucosubstance of *T. pacificus*, mucin was easily isolated by solubilization with 0.4 M NaOH, followed by precipitation with 50 % ethanol in the presence of 0.2 M NaCl. About 170 mg of the nidamental mucin was obtained from 3 g (wet weight) of the mucosubstance. First, the mucin was analyzed for chemical constituents. The results are given in Table 1 in terms of residues per 1,000 total amino acid and sugar residues, confirming that the mucin has the compositional feature as a sulfated glycoprotein. It comprised 16.4 % protein, 80.3

Table 1. Chemical compositions of nidamental mucin and its protease-resistant fragment from Japanese common squid.

	Nidamental mucin	Protease-resistant fragment
Amino acid		
Asp	9	4
Thr	93	96
Ser	7	8
Glu	7	3
Pro	51	54
Gly	8	6
Ala	5	3
Cys	2	0
Val	9	7
Met	1	0
Ile	45	45
Leu	5	2
Tyr	2	1
Phe	4	0
His	5	6
Lys	6	2
Arg	3	0
Total	(262)	(237)
Neutral sugar		
Gal	200	197
Man	7	8
Glc	20	25
Xyl	5	9
Fuc	141	176
Rha	7	11
4-*O*-MeGlc	43	45
Total	(423)	(471)
Amino sugar		
GlcNAc	120	116
GalNAc	130	136
4-*O*-MeGlcNAc	65	40
Total	(315)	(292)
Total	1000	1000
Ester sulfate	56	68

Values are given in residues/1,000 total amino acid and sugar residues.

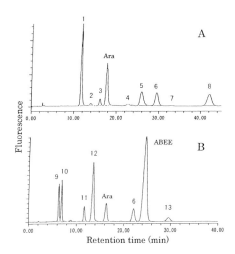

Figure 1. Chromatographic separation of ABEE-derivatized monosaccharides from nidamental mucin. The ABEE-derivatized monosaccharides were eluted from a Honenpak C18 column at 45 ℃ with (A) 0.2 M potassium borate buffer (pH 8.9) containing 6 % acetonitrile and (B) 0.02 % TFA containing 10 % acetonitrile at a flow rate of 1.0 mL/min. L-Arabinose (Ara) was used as an internal standard. The identities are as followes: 1, Gal ; 2, Man ; 3, Glc ; 4, Xyl ; 5, GlcNAc ; 6, Fuc ; 7, Rha ; 8, GalNAc ; 9, GlcN ; 10, GalN ; 11, 4-*O*-MeGlcN ; 12, Gal+Glc ; 13, 4-*O*-MeGlc.

% sugar and 3.3 % ester sulfate by weight and was characterized by the high contents of threonine, *N*-acetylglucosamine, *N*-acetylgalactosamine, fucose and galactose. Approximately 72 % of the protein backbone was composed of three amino acids, threonine, proline and isoleucine, which were present in the molar ratio of approximately 2:1:1. In contrast to serine residues, threonine residues accounted for more than one-third of the total amino acids and thus were assumed to be the main candidate for protein-sugar linkages in the mucin. The chromatographic separation of ABEE-derivatized monosaccharides from the mucin is given in Fig. 1. Of particular interest is the existence of two methylated monosaccharides and complete separation of them from other monosaccharides is achieved by using 0.02 % TFA containing 10 % acetonitrile as the eluent (Fig. 1B) ; peaks 11 and 13 are identified as 4-*O*-methylglucosamine and 4-*O*-methylglucose, respectively. 4-*O*-methylglucosamine as well as glucosamine and galactosamine was naturally present as *N*-acetylated form, because the equimolar amounts of acetic acid and amino sugars were obtained

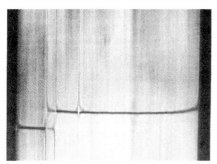

Figure 2. Sedimentation velocity pattern of nidamental mucin at 28 ℃. The mucin was dissolved in 0.02 M sodium phosphate buffer (pH 7.2) containing 0.1 M NaCl at a concentration of 0.030 %. The photograph was taken by the schlieren method at 13 min after the speed of 50,000 rpm was reached.

Table 2. Macromolecular properties of nidamental mucin and its protease-resistant fragment.

Mucin[*1]	[η](dL/g)	$S^0_{20,w}(S)$	Molecular mass (kDa)
Nidamental mucin	12.3	16.9	2,600
Protease-resistant fragment	14.4	14.3	2,200

[*1] Alkali-solubilized mucin.

after acid hydrolysis of the mucin (data not shown). The nidamental mucin is found to be relatively rich in both methylated monosaccharides ; 65 residues/1,000 for. N-acetyl-4-O-methylglucosamine and 43 residues/1,000 for 4-O-methylglucose (Table1).

The molecular shape and molecular weight of nidamental mucin were estimated by both sedimentation velocity analysis and viscosity measurement. As given in Fig. 2, a single, sharp peak having a sedimentation constant of 16.9 S is observed by ultracentrifugal analysis of the mucin solution at 28 ℃. In the same temperature, it had a high intrinsic viscosity of 12.3 dL/g, which is presumably a minimum value because of the relatively high shear gradient (1,000 sec^{-1}) of the viscometer used. From the equation of Simha[14], the axial ratio of mucin molecules in solution as a prolate ellipsoid of evolution was calculated to be about 193. From these results, we calculated a molecular mass of the mucin to be about 2,600 kDa by using the Scheraga-Mandelkern equation[15] with a value of 3.42 x 10^6 for β.

The mucin was very resistant to proteolytic action and a large fragment was obtained by precipitaion with 60 % ethanol after actinase E digestion of the mucin at 50 ℃ for 17 h. The yield was about 86 %. As shown in Tables 1 and 2, there is no marked change in its chemical composition and macromolecular structure, respectively, before and after protease digestion of the mucin. Formation of the large, protease-resistant fragment having the extremely high contents of threonine, proline and isoleucine (Table 1) and a molecular mass of 2,200 kDa (Table 2) imply that molecular ends of the mucin are relatively poor in these amino acids and are removed by actinase E. Incidentally, the sum of threonine, proline and isoleucine is increased by the protease digestion from 72 % to 82 % of the total amino acids (Table 1).

From these combined results, the alkali-solubilized mucin of *T. pacificus* was found to be similar in its molecular properties to that of *I. argentinus*.[3] The intact mucin in the mucosubstance appears to be associated with other proteins to form a huge, complex structure,[4] and was solubilized by the mild alkaline treatment. However, solubilization mechanism of the intact mucin remains to be solved.

Properties of a Major Oligosaccharide Released by Hydrazinolysis of Mucin

The nidamental mucin contains the methylated monosaccharides as described above and thus the structure of O-glycosidically linked sugar chains containing these unusual constituents is very interesting. To isolate such sugar chain, the mucin was treated with 0.5 M NaOH at 25 ℃ for 24 h. However, there was no sugar chain released by β-elimination reaction. This was confirmed by the finding that no significant change in the threonine content of mucin was observed before and after the alkaline

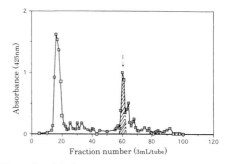

Figure 3. Gel filtration of oligosaccharides released by hydrazinolysis of nidamental mucin. The oligosaccharides were eluted from a Bio-Gel P4 (Pharmacia) column at 55 ℃ with distilled water at a flow rate of 12.4 mL/h. The effluent was monitored by the orcinol-sulfuric acid method. A major neutral oligosaccharide I was obtained.

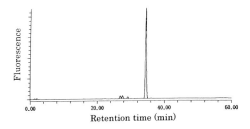

Figure 4. Chromatographic pattern of a major ABOE-derivatized oligosaccharide I released by hydrazinolysis of nidamental mucin. The oligosaccharide was eluted from a Honenpak C18 column at 45 ℃ with two solvents, A and B, at a flow rate of 1.0 mL/min as described in "Materials and Methods".

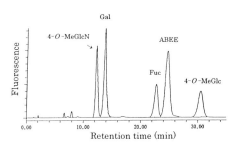

Figure 5. Chromatographic separation of ABEE-derivatized monosaccharides from oligosaccharide I. The HPLC conditions were the same as those described for (B) in the legend of Fig. 1.

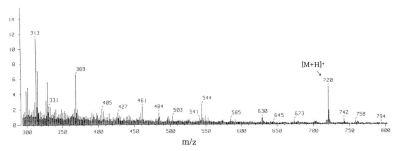

Figure 6. Positive mode FAB mass spectrum of oligosaccharide I.

treatment. Moreover, the modified Morgan-Elson reaction was negative for the mucin, suggesting the lack of terminal 3-O-substituted N-acetylgalactosamine which is present in all vertebrate mucins so far examined.[9] Then, mild hydrazinolysis with anhydrous hydrazine at 60 ℃ for 50 h was applied to this squid mucin. As shown in Fig. 3, the sugar chains released by hydrazinolysis, followed by N-acetylation of free amino groups, are separated into several oligosaccharides on a Bio-Gel P4 column, although the largest peak at the fractions 17 - 20 was found to contain mainly the mucin resistant to this mild hydrazionolysis. A major neutral oligosaccharide I at the fractions 59 - 61 was further analyzed by reversed-phase HPLC of its ABOE-derivative on a Honenpak C18 column ; Fig. 4 reveals the presence of a single peak together with negligibly small ones. The ABEE-derivatized monosaccharides of I are found by HPLC with the TFA solution as the eluent to consist of the equimolar amounts of galactose, fucose, 4-O-methylglucose and 4-O-methylglucosamine (Fig. 5). The occurrence of galactose in I was confirmed by HPLC with the borate buffer as the eluent, because ABEE-galactose and ABEE-glucose are not separated by HPLC with the TFA solution (Fig. 1B). On the other hand, the FAB mass spectrum of I reveals a peak at m/z 720 for [M+H]$^+$, indicating that its molecular weight is 719 (Fig. 6). These results imply that the oligosaccharide I is a tetrasaccharide containing 1 mol each of galactose, fucose, 4-O-methylglucose and N-acetyl-4-O-methylglucosamine. The lack of N-acetylgalactosamine in this tetrasaccharide is in contrast with the observation that mucin-type oligosaccharides are generally linked to threonine and/or serine residues in the protein backbone through N-acetylgalactosamine.[16] Thus, a possibility that the tetrasaccharide of nidamental mucin is linked to threonine residues through another monosaccharide than N-acetylgalactosamine is not excluded ; for instance, an oligosaccharide O-glycosidically linked to a serine residue through fucose was found in human factor IX.[17] Further structural studies on the isolated tetrasaccharide, which could not be released by β - elimination, are needed for elucidation of a protein-sugar linkage in the nidamental mucin.

Acknowledgments

We wish to thank Dr. Nishimura T. (Forestry and Forest Products Research Institute, Tsukuba) for a sample of 4-O-methylglucose and Dr. Murakami M. (University of Tokyo) for FAB mass spectrometry.

References

1) DURWARD R D, VESSEY E, O'DOR R K and AMARATUNGA T : Reproduction in the squid, *Illex illecebrosus* : First observation in captivity and implications for the life cycle. ICNAF Sel. Papers 6, 7-13 (1980).
2) BOWER J R and SAKURAI Y : Laboratory observations on *Todarodes pacificus* (Cephalopoda: Ommastrephidae) egg masses. Amer. Malac. Bull. 13, 65-71 (1996).
3) KIMURA S, SUGIURA Y, MIZUNO H, KATO N and HANAOKA Y : Occurrence of a mucin-type glycoprotein in nidamental gland mucosubstance from the squid *Illex argentinus*. Fish. Sci. 60,193-197 (1994).
4) SUGIURA Y and KIMURA S : Nidamental gland mucosubstance from the squid *Illex argentinus* : Salt-soluble component as a mucin complex. Fish. Sci. 61, 1009-1011 (1995).
5) KIMURA S, TSUNOO T, MATSUDA H, MURAKAMI M and YAMAGUCHI K : Occurrence of 4-O-methyl-D-glucosamine in squid mucin. Fish. Sci. 65, 325-326 (1999).
6) IMANARI T, ARAKAWA Y and TAMURA Z : Gas chromatographic analysis of aldose. Chem. Pharm. Bull. 17, 1967-1969 (1969).
7) YASUNO S, MURATA T, KOKUBO K, YAMAGUCHI T and KAMEI M : Two-mode analysis by high-performance liquid chromatography of p-aminobenzoic ethyl ester-derivatized monosaccharides. Biosci. Biotech. Biochem. 61, 1944-1946 (1997).
8) NISHIMURA T, ISHIHARA M, ISHII T and KATO A : Structure of neutral branched xylooligosaccharides produced by xylanase from *in situ* reduced hardwood xylan. Carbohydr. Res. 308, 471-476 (1988).
9) BHAVANANDAN V P, SHEYKHNAZARI M and DEVARAJ H : Colorimetric determination of N-acetylhexosamine-terminating O-glycosidically linked saccharides in mucins and glycoproteins. Anal Biochem. 188, 142-148 (1990).
10) TERHO T T and HARTIALA K : Method for determination of the sulfate content of glycosaminoglycans. Anal. Biochem. 41, 471-476 (1971).
11) SCAWEN M S and ALLEN A : The action of proteolytic enzymes on the glycoprotein from pig gastric mucus. Biochem. J. 163, 363-368 (1977).
12) KURAYA N and HASE S : Release of O-linked sugar chains from glycoproteins with anhydrous hydrazine and pyridylamination of the sugar chains with improved reaction conditions. J. Biochem. 112, 122-126 (1992).
13) WINZLER R J : Determination of serum glycoproteins, in : 「Methods of Biochemical Analysis」(ed. by GLICK D), Vol. II. Interscience, New York, 1955, pp. 279-311.
14) SIMHA R : The influence of Brownian movement on the viscosity of solutions. J. Phys. Chem. 44, 25-34 (1940).
15) SCHERAGA H A and MANDELKERN L : Consideration of hydrodynamic properties of proteins. J. Am. Chem. Soc. 75, 179-184 (1953).
16) STROUS G J and DEKKER J : Mucin-Type Glycoproteins. Cri. Rev. Biochem. Mol. Biol. 27, 57-92 (1992).
17) NISHIMURA H, TAKAO T, HASE S, SHIMONISHI Y and IWANAGA S : Human factor IX has a tetrasaccharide O-glycosidically linked to serine 61 through fucose residues. J. Biol. Chem. 267, 17520-17525 (1992).

索 引

各語は五十音順に配列したが，欧文で始まる語は一括して末尾においた．なお，物質名の接頭語はこれを無視した．（例，N-アセチル化は ［あ］ の項に）

［あ］

アオリイカ　11
　―包卵腺粘質物　13, 14
アカイカ　V, 11
　―科　9, 10, 11, 12
　―包卵腺粘質物　13, 14
N-アセチル化　13, 19, 35, 41, 42
N-アセチルガラクトサミン　3, 4, 25
N-アセチルグルコサミン　4, 55
N-アセチル-4-O-メチルグルコサミン　19, 36
　―の系統名　19
アテロコラーゲン　67, 68, 69
アテロムチン　29, 40, 66, 68, 69
p-アミノ安息香酸エチルエステル　35
p-アミノ安息香酸オクチルエステル　41
アミノ酸　6, 12, 23, 25, 28, 40, 55, 57
　―残基　6
　―の反復配列　4, 6, 41
　―配列　6
アミノ糖　11, 15, 19, 35, 69
アミノ糖 X ［＝4-O-メチルグルコサミン］　11, 12
　―の拡張二次元 NMR 分析　18
　―の構造　15, 19
　―の単離　15, 16
　―の分子式　18
　―の分子量　15

　―の FAB-MS スペクトル　17
　―の ^{1}H および ^{13}C NMR データ　16, 18
　―の ^{1}H-^{1}H COSY スペクトル　21
　―の HMBC スペクトル　22
アルゼンチンイレックス　V, 9, 11, 23
　―包卵腺粘質物　VI, 9, 15, 23
　―包卵腺ムチン ［→ 包卵腺ムチンの項も参照］　VI, 23
イカ　V, 9
　―の漁獲高　V
　―の性成熟度　9
　―の包卵腺粘質物　9
イソロイシン　23, 28, 36, 40, 57
遺伝子群　5
エクアトリアル結合　25
エタノール沈殿　7, 23, 30, 36, 55
塩溶性成分　30 ～ 34
　―のアミノ酸・糖組成　31, 32
　―の固有粘度　33
　―の沈降速度図　33, 34
　―の沈降定数　33
　―の分子量　33
　―のムチン　32, 33, 34
　―の SDS-ゲル電気泳動図　32
塩溶性ムチン複合体　29, 34

［か］

核磁気共鳴法　15

索引 **85**

ガスクロマトグラフ　11
ガラクトース　4, 42
還元剤　7, 14
希アルカリ　4, 65
　―可溶化(法)　65, 66
　―処理　5, 23, 41, 55, 57, 65
逆相 HPLC　35
極限沈降係数［＝沈降定数］　33
巨大卵塊［→スルメイカの項も参照］
　　V
グアニジン塩酸　7, 55
N-グリコシド型糖鎖　4
O-グリコシド型糖鎖　3, 4
O-グリコシド結合　3, 4, 5
　GalNAc-Thr 間の―　25, 26
系統発生　5
化粧品素材　66
ゲノム配列　6
ゲル
　―形成　6
　―形成能　35
　―状膜　23
　―電気泳動　14
　―濾過　7, 41, 42
コアタンパク質　3, 4, 5, 28, 33, 36, 47, 57
高速液体クロマトグラフ　11
高速原子衝撃-質量分析法　15
コラーゲン　10, 23, 65, 67, 68

[さ]
細繊維　55
　―のアミノ酸組成　61
　―のアミノ酸・糖組成　56

　―の起源　60
　―の SDS-ゲル電気泳動図　60, 62
桜井泰憲　V, 53
産卵行動　VI
　―の記録　53
　―の模式図　53, 54
シアル酸　4, 5, 7, 25
上皮細胞　3, 5
ジンドウイカ科　10, 11, 12
スルメイカ　V, 11, 35
　―の解剖模式図　9, 10
　―の巨大卵塊　III, V, VI, 53, 64
　―の産卵行動　VI, 53, 64
　―の包卵腺　9, 10, 53
　―の包卵腺粘質物　13, 14
　―の包卵腺ムチン　35, 36
　―の輸卵管　10
　―の輸卵管腺粘質物　12, 13
　―類　V, 10, 53
生体防御物質　3
ゼリー(状)物質　53
セリン　3, 4

[た]
単糖　3, 6, 25, 41, 60
タンパク質　VI, 4, 5, 6
　―組成　10, 14
　―分解酵素［＝プロテアーゼ］　5, 26
中性糖　11, 35, 41, 55, 60
沈降速度測定　36
纏卵腺［＝包卵腺］　9, 53
糖
　―アルコール　11

―残基　　6
　　　―質　　11, 12, 28, 33, 36, 57
　　　―分析法　　35
糖鎖　4, 6
　　　―の還元末端　　4
　　　―の切り出し　　4, 41
　　　―のゲル濾過　　41, 42
　　オリゴ―　　3, 5, 6
　　ムチン型―　　3
糖鎖 I　　41
　　　―の拡張二次元 NMR 分析　　46
　　　―の還元末端　　47
　　　―の単離　　41
　　　―の薄層クロマトグラフ　　41, 43
　　　―の分枝構造　　47
　　　―の分子量　　42
　　　―のメチル化糖　　48
　　　―の ABEE 化単糖　　44
　　　―の ABOE 化　　41
　　　―の COSY スペクトル　　46, 49
　　　―の FAB-MS（スペクトル）　　42, 44
　　　―の ^1H および ^{13}C NMR データ
　　　　　45, 46
　　　―の HMBC スペクトル　　46, 50
　　　―の ROESY スペクトル　　51, 46
糖タンパク質　　3, 4, 5, 7, 25, 29, 47, 55, 63
　　　―ムチン　　Ⅵ , 3
ドメイン　　4, 5, 6
　　細胞接着―　　5
　　システイン含有―　　5
　　膜貫通―　　5
トリフルオロ酢酸　　11, 35
トレオニン　　3, 4, 23, 25, 28, 36, 40, 57

［な］
尿素　　7, 14
粘液　　7
　　　―ゲル層　　3
　　　―状物質　　53
　　　―性　　3
粘質物［→ 包卵腺粘質物］
粘膜　　3, 7
粘度測定　　36, 68

［は］
ヒドラジン分解　　41
フコース　　4, 42
プロテアーゼ　　26, 29
　　　―消化　　26, 40
プロテアーゼ抵抗性断片　　27, 40
　　　―の化学組成　　24, 28, 38
　　　―の固有粘度　　29, 40
　　　―の沈降速度図　　28, 30
　　　―の沈降定数　　40
　　　―の分子量　　40
プロリン　　3, 23, 28, 36, 40, 57
ペプチド結合　　5, 69
包卵腺　　V, 9, 10, 11, 53
包卵腺粘質物　　9 〜 15, 27, 55, 57, 63, 64
　　　―のアミノ酸組成　　61
　　　―のアミノ酸・糖組成　　11, 13
　　　―の一般組成　　11
　　　―の塩溶性成分［→ 塩溶性成分の項も参照］　　34
　　　―の可溶化ムチン　　23, 55
　　　―の可溶性成分　　30, 31, 32
　　　―の希アルカリ処理　　65

索 引　87

―の水溶性ムチン　55, 58
―のタンパク質組成　14
―の不溶性成分　60
―の分画　29, 55
―の未知アミノ糖［→ アミノ糖 X の項も参照］　15
―の SDS-ゲル電気泳動図　14, 27, 62
包卵腺ムチン［→ マリンムチンの項も参照］　VI, 23, 36, 65, 66
　―のアミノ酸・糖組成　31, 34
　―の化学組成　23, 24, 25, 28, 36, 38
　―の還元粘度　25, 28
　―の希アルカリ可溶化　23, 65
　―の固有粘度　33, 39, 40, 59
　―の主要糖鎖 I　41
　―の商品名［＝マリンムチン］　66
　―の製造　66
　―の赤外吸収（スペクトル）　25, 26
　―の単離（法）　23, 33, 36, 65
　―の沈降係数　28, 39
　―の沈降速度図　28, 39
　―の沈降定数　39, 40, 59
　―のプロテアーゼ消化　26, 40, 66
　―のプロテアーゼ抵抗性断片　27, 40
　―の分子形　40
　―の分子中央部　29, 40
　―の分子末端部　29, 41
　―の分子量　40, 59
　―の利用　65
　―の SDS-ゲル電気泳動図　25, 27, 28
ポリペプチド鎖　3

［ま］
マトリックス　15, 17, 44
マリンムチン［＝化粧品素材］　66, 69
　―の安全性　67
　―の吸湿性　69
　―の抗原性　67
　―の耐熱性　69
　―の粘性　68
ムコ多糖類　5
ムチン　3, 5
　―遺伝子（群）　5, 6
　―型糖鎖　3, 4
　―型糖鎖の結合様式　4
　―の可溶化　7, 23
　―の系統発生　6
　―のゲル濾過　7
　―のコアタンパク質　3, 4, 5
　―の精製　7
　―の生理機能　65
　―の多量体　5
　―の単離　7
　―の定義　3
　―の模式図　4
　―の利用　65
　―複合体　34, 63
　―分子　5, 29
魚類―　7
ゲル形成―　5, 6
ゲル形成―の起源　6
シアロ―　4
スルホ―　4, 25
脊椎動物―　5, 25, 46, 65, 66
非ゲル形成―　7
分泌型―　3, 5, 6, 7

哺乳類―　7
膜結合型―　3, 5, 6, 7
メチル化(単)糖　19, 36, 48
4-O-メチル(-D-)グルコサミン［＝アミノ糖 X］　11, 15, 19, 20, 42
　―の系統名　19
4-O-メチルグルコース　19, 35, 36, 42

―の固有粘度　58, 59
―の沈降速度図　59
―の沈降定数　58, 59
―の分子形　59
―の分子量　58, 59
硫酸イオン　25, 39
硫酸基　4, 5, 7, 23, 25, 28, 36, 55

[や]

輸卵管腺　V, 53
輸卵管腺粘質物　53, 60, 64
　―のアミノ酸組成　61
　―のアミノ酸・糖組成　13
　―の SDS-ゲル電気泳動図　62

[ら]

卵塊　V
　―形成　VI, 55, 63
　―の表層ゲル状膜　53, 55
卵塊内部ゼリー　53, 54, 55, 64
　―のアミノ酸組成　60, 61
　―の細繊維［→ 細繊維の項も参照］　60, 63, 64
　―のドリップ　55, 60, 61
　―のムチン複合体　60
卵塊膜　5, 53, 54, 55, 60, 63
　―の化学組成　55, 56
　―の可溶化ムチン　57
　―の起源　57
　―のムチン複合体　55, 57
　―の ABEE 化中性糖　57
　―の SDS-ゲル電気泳動図　55, 58
卵塊膜ムチン　57
　―の化学組成　56, 58

[欧文]

ABEE　35
　―化単糖　37
　―化中性糖　57
ABOE 化糖鎖 I　41
atelomucin　29
α-アノマー　16, 18, 21, 22
β-アノマー　16, 18
β 離脱反応　4, 25, 41, 65
C 末端　5
^{13}C NMR　15〜18, 45, 46
COSY　46, 49
D 型　18
DEPT　17
DMSO　45, 46
DTT　14
ELISA 法　67
FAB-MS　15, 17
Fuc　42, 44, 46, 47
Gal　42, 44, 46, 47
GalNAc　3, 4, 25, 47
^{1}H および ^{13}C NMR データ　16, 18, 45, 46
^{1}H-^{1}H COSY スペクトル　18, 19, 20
HMBC　18, 19, 22, 46, 50
^{1}H NMR　15, 16, 18, 45, 46

索　引

HPLC　　11, 35, 41
Ile　　40, 41
4-*O*-MeGlc　　42, 46, 47
4-*O*-MeGlcN　　42, 44
4-*O*-MeGlcNAc　　42, 46, 47
MUC　　5
MUC 遺伝子ファミリー　　5
mucin　　3
mucosa　　3
N 末端　　5

NMR　　15
Pro　　40, 46
PTS ドメイン　　4, 5, 6
ROESY　　46, 51
SDS　　14
SO_4^{2-}　　25
S-S 結合　　5, 15, 25, 26, 36, 57
TFA　　11, 35
Thr　　25, 40, 41

おわりに

　ムチンに出会ってから25年ほどになる．私の主要な研究テーマは海洋生物に由来する繊維性コラーゲン(海洋性コラーゲン)であり，ムチンは当初，包卵腺粘質物の依頼分析から派生した糖タンパク質の一種に過ぎなかった．しかし，依頼分析が終了したムチンと私の縁はコラーゲンとの関係で，その後も途絶えることが無かった．

　コラーゲンの主要な機能の一つは動物の体表面(皮膚)を外界からの異物や刺激から物理的に保護することである．一方，ムチンはすでに述べたように傷つき易い消化管，気管などの体内表面を保護する粘膜の主要な糖タンパク質で，同じ糖タンパク質に属するコラーゲンとは多くの共通点をもつ兄弟のような関係にある．両者は共に動物の体や細胞に優しい生体防御物質と考えられる．このような観点からムチンの研究を進めていくと，次々と新しい展開があり，予想外の実り多い成果が得られた．コラーゲン繊維を構成する著しく細長い剛体棒状の分子は粘性の高い水溶液を生じる．長年にわたるコラーゲン分子の研究で培った経験は粘液性の包卵腺ムチンの研究に大いに役立ち，その礎になったといえよう．

　これまでにイカのムチンに関する研究は私の知る限り他に無く，得られた知見を要約すると次のとおりである．

1) スルメイカ類の包卵腺粘質物を試料として，粘性の原因物質は分子量200万を超える巨大な糖タンパク質ムチンであることを立証し，その簡便な調製法(希アルカリ可溶化法)を確立した．

2）ムチンから新規の2種のメチル化単糖を分離して構造を決定した．
3）これらメチル化単糖を含む4糖からなる主要な糖鎖Iは脊椎動物のムチン糖鎖と異なり，還元末端のガラクトースを介してコアタンパク質にO-グリコシド結合した枝分かれの多い特異な構造をもっている．
4）ムチンは他のタンパク質と複合体を形成し，スルメイカが生み出す直径40〜100 cmの巨大な球形卵塊の支持組織となり，表層の卵塊膜および成熟卵を囲む卵塊内部ゼリーの細繊維を構成している．
5）次いで海洋生物ムチンとして初めて，包卵腺ムチンの商品化に成功し，化粧品素材への道を拓いた．

　こうして，ムチンの研究は多くの共同研究者と予期せぬ幸運とに恵まれた楽しい思い出となった．ところで，スルメイカ類はどのような方法で包卵腺から粘質物を体外に放出し，数分の内に，巨大な卵塊の表層ゲル状膜と内部ゼリーを形成するのであろうか．この巧妙な仕組みが解明されることを期待したい．
　最後に，貴重なスルメイカの包卵腺と卵塊を提供していただいた桜井泰憲博士（北海道大学名誉教授）に厚くお礼を申し上げる．本研究に対しては平成8〜9年度科学研究費補助金（基盤研究(B)(2)）および(株)紀文食品の奨学寄付金を受け，また五曜書房の日吉尚孝社長には出版に際して大変お世話になった．ここに感謝の意を表する．

2016年9月1日　　　　　　　　　　　　　　　　　　　　　　　　著者

著者略歴

木村　茂（きむら　しげる）
1940年東京向島生まれ．東京水産大学製造学科を卒業の後，日本皮革（株）を経て，東京水産大学（現，東京海洋大学）に永く勤務する．現在，東京海洋大学名誉教授．著書に「海洋性コラーゲンを探る」（五曜書房）がある．

スルメイカ類のムチン　―巨大卵塊の形成―

2016年9月28日　初版第1刷

著　者　　木村　茂
発行人　　日吉尚孝
発行所　　株式会社五曜書房
　　　　　〒101-0065　東京都千代田区西神田 2-4-1　東方学会本館3F
　　　　　電話（03）3265-0431
　　　　　振替 00130-6-188011
発売元　　株式会社星雲社
印刷・製本　株式会社太平印刷社
ISBN978-4-434-22472-0
定価はカバーに表示してあります．落丁・乱丁本はお取替えいたします．